Praise for *A Walk Among Heroes*

"*A Walk Among Heroes* captures the essence of Arlington, not just as a resting place for warriors, but as a monument to the enduring spirit of those who stood behind them—the spouses. These unsung heroes are given their rightful place in this narrative. This book isn't just a reflection on the fallen; it's an homage to the resilience and quiet strength of the families who have been the backbone of our forces. As someone who has felt the heavy blend of love and grief that Arlington evokes, I see in these pages a tribute to the full spectrum of sacrifice and loyalty that defines the military family. The book is a powerful reminder that heroism comes in many forms, often unseen, but always felt."

—Britt Slabinski, Command Master Chief (SEAL), USN (Retired), Medal of Honor recipient

"Zelinski does a masterful job weaving his photos and stories into something extraordinary. The stories about the two officers who lost their lives while serving at the National Reconnaissance Office and Gary's way of honoring their spouses' service reveal the toll they also paid for our freedom. As a former Deputy Director of the NRO, I know that these officers and many others resting at Arlington are our nation's unsung heroes who served while guarding our nation's secrets."

—Stephen T. Denker, Major General, USAF (Retired)

"*A Walk Among Heroes* offers a poignant exploration of both renowned and overlooked figures buried at Arlington National Cemetery. Zelinski's narrative seamlessly blends personal anecdotes, historical accounts, and reflective insights, immersing readers in the lives of those who shaped our collective past. It is a moving tribute to the enduring power of human connection and the profound impact of ordinary individuals on history. With compelling storytelling and rich historical detail, the book invites readers on a transformative journey of remembrance and reverence."

—Sandra Miller Linhart, award-winning author of *Daddy's Boots* and *Frozen Tears*

"Gary's take on this revered ground is refreshingly different from any other book or article I have read. His decision to use black-and-white photos, which he took, adds a stark, almost haunting garnish to the fascinating story he weaves. No matter how much you've read on the subject or how many visits you have made, I urge you to read this book; you will see Arlington in a new and interesting light."

—George D. Bremer, Jr., Colonel, USAF (Retired)

"I was so absorbed in the writings I could not put the book down until reaching the end. I was fascinated, I was angry, I was excited, I was spellbound! Gary has the knack—his enviable skills in humanizing those stone monuments in Section 60 and Arlington as a whole were magical! His short, fierce human tributes make the reader want the stories on everyone at Arlington. I strongly recommend this book be read by all Americans!"

—Roald J. Moyers, Chief Warrant Officer, USA (Retired); Master, SIGINT Technical Expert, National Security Agency, National Cryptologic University; and Professionalized Signals Collection Officer

"Gary Zelinski has crafted a thoughtful, well-researched, loving look at a select few of the veterans buried at Arlington National Cemetery. Covering those who are well known and a few you may never have heard of, *A Walk Among Heroes* serves as a testament to all those buried in our national cemeteries. An inspiring read on Memorial Day or any day."

—Greg Elliot, UCLA instructor and screenwriter

"I've known about Arlington National Cemetery my entire adult life but never to the extent as told by Gary Zelinski in his wonderful book, *A Walk Among Heroes*. I loved his stories of some of the notables laid to rest in this hallowed ground. I couldn't put it down."

—Nancy Panko, award-winning author of *Guiding Missal* and *Sheltering Angels*

"*A Walk Among Heroes* offers a fitting tribute to the greatest boxer of all time, Joe Lewis. 'The Sweet Science' chapter also captures the history and beauty of a sport I've dedicated my life to. The book is as inspiring as it is educational. Well done!"

—Fausto De La Torre, President, Golden Gloves of California

“Thank you for including the story about Micheal Spann and General Donovan. You honored their memory and gave life to the service I dedicated my life to. Some of the people you mentioned are well known to me, but you found a way to shine new light on many familiar heroes.”

—Dan K., retired senior CIA officer

“*A Walk Among Heroes* tells the stories of soldiers, airmen, Marines, wives, a boxer, a musician, presidents, and astronauts, all buried at Arlington. At the end of May, you will look at the flowers in bloom and think of the men and women who give meaning to Memorial Day.”

—William Rose, retired Special Agent, FBI, and Specialist, United States Army 3rd Infantry Division

“This book resonated with me profoundly because I worked at the cemetery as an architect during the restoration of the north wing of Arlington House. It was impossible to be there without a sense of reverence for that hallowed ground and those honored there. *A Walk Among Heroes* beautifully expresses that reverence and offers a deeper understanding of why Arlington National Cemetery is a timeless expression of America's search for our ‘better angels.’”

—S. Elizabeth Sasser, AIA, Principal, Quid Tum Historic Structures Consulting, and retired Restoration Archeologist, National Park Service

“Zelinski's captivating photos and words transport the reader from battlefield to graveside to family picnics to unforgettably traumatic yet fiercely proud moments of our country's history. Traveling through time and Arlington Cemetery is an experience that will stay with you long after reading this unforgettable book.”

—Dr. Jane Hayes, Professor Emerita, Computer Science Department, University of Kentucky

“Zelinski gives the reader a deeper and more meaningful understanding of the some of the heroes buried at Arlington National Cemetery. When you read these pages, you'll feel like you are walking the lanes of Arlington as you learn, laugh, and cry.”

—Bob Jordan, Detective, Gilbert Police Department Street Crimes Unit

"Heartfelt and thoughtfully written, the author's words touched my soul. Some stories made me cry but also made me proud to be an American. As a former teacher and administrator for Los Angeles Unified School District, I resolutely advocate for this book to be used in middle and high school. I know students would greatly benefit from the history and personal stories in this book!"

—Carmen Gilmore, President, Canyon Country Optimist Club, and retired teacher and administrator, Los Angeles Unified School District

"The book is sure to trigger some nostalgia. I was particularly drawn to the Tomb of the Unknowns. In my earlier years, my social circle included the men of the Old Guard. I always got a kick out of watching them transition seamlessly from a typical twenty-year-old guy to the dignified soldier guarding the tomb or carrying the colors behind the president."

—Julie Tucker, Vice President for Client Relations, The Ryding Company, and Yeoman Chief Petty Officer (Air Warfare), USN (Retired)

"*A Walk Among Heroes* is a thoroughly researched tribute to a few American veterans buried at Arlington National Cemetery. Gary shines a light on some people and their families who would otherwise go unnoticed by the casual visitor. This book sparks a curiosity to discover more about other people interred in US national cemeteries. It is a must-read for anyone who knows someone who has served their country."

—Billy Marchal, Chief Master Sergeant, USAF (Retired), Air Force Functional Manager, Pentagon; C4I Air Force Operational Staff; NATO/UN Senior Enlisted Advisor/Joint Forces

"As an Army private stationed at Fort McNair in Washington DC, I often visited Arlington National Cemetery. I did, however, see only the forest—never the trees. *A Walk Among Heroes* has changed that, opening my eyes to the wonder that each grave has a story of the life and death of that hero and that family's sacrifices. Zelinski transforms the ordered coldness of a cemetery into meticulously researched, patriotic, yet warm, thoughtful, and sentimental accounts of American heroes."

—Sue Rushford, Principal, Sue Rushford Editorial Services, and Army veteran

A WALK AMONG HEROES

A WALK AMONG HEROES

SEARCHING FOR AMERICA'S BETTER ANGELS

GARY B. ZELINSKI

Disclaimer: DoD or IC approval of this nonofficial manuscript does not imply endorsement or authentication. The opinions and views expressed herein are those of the author and do not reflect the opinions or views of the USG.

Author's note: Due to the sensitive nature of many of my military assignments, the names of some colleagues have been removed or changed. Some of my personal recollections have also been modified. My friends know who they are and what we did.

Blue Hexagon Publishing
P.O. Box 971
Camarillo, CA 93011
http://www.bluehexagonpublishing.com

Mama Joe's Tablecloth, courtesy of the Smithsonian National Air and Space Museum, photograph by Jonathan R. Barrett.

Quantity sales. Special discounts are available on quantity purchases by corporations, associations, and others. For details, contact the "Special Sales Department" at the address above.

Orders by US trade bookstores and wholesalers. Please contact BCH: (800) 431-1579 or visit www.bookch.com for details.

Printed in the United States of America

Cataloging-in-Publication Data

Names: Zelinski, Gary B., author.
Title: A walk among heroes : searching for America's better angels / Gary B. Zelinski.
Description: Includes index. | Camarillo, CA: Blue Hexagon Publishing, 2025.
Identifiers: LCCN: 2024926626 | ISBN: 978-1-963954-10-4
Subjects: LCSH Military biography. | United States. Army—Biography. | United States. Navy—Biography. | United States. Air Force—Biography. | United States. Marine Corps.—Biography. | Arlington National Cemetery (Arlington, Va.) | Arlington National Cemetery (Arlington, Va.)—History. | BISAC BIOGRAPHY & AUTOBIOGRAPHY / Military | BIOGRAPHY & AUTOBIOGRAPHY / Historical
Classification: LCC F234.A7 .Z45 2025 | DDC 917.55295—dc23

First Edition

Cover designer: Jenny Kimura

For those who seek the better angels of our nature.

CONTENTS

Monuments

Morale

Ad Astra Per Aspera

INTRODUCTION

Arlington National Cemetery

A casual walk at Arlington National Cemetery on a warm spring day turned into a decades-long journey—a search for America's better angels. I was planning a picture book, just a bunch of black-and-white photographs with limited cropping and editing, preserved on acetate, not in the cloud. When I enlisted in the Air Force long ago, my first job was as a photographer. I learned the craft of taking pictures when nothing was automatic or digital. Every setting was purposeful and deliberate. I learned to create photographs, not just take pictures.

You can't see all your heroes in one walk at Arlington National Cemetery. The place is enormous—639 acres with over four hundred thousand people enshrined on its gently rolling hills. As one walk turned into two, then more, the years passed. My wife and I moved away. Life moves on, but Arlington remained. I had too few photographs—not enough for a book, just enough to continue the dream. With no Arlington lanes to walk, I had time to learn about some of the famous heroes buried: Doolittle, Donovan, and Murphy. I explored the selfless deeds of Basilone and Levitow. I relived painful memories of the struggle for civil rights with Evers.

At Arlington, it is said "Every stone tells a story." What follows are a few accounts, my attempt to move beyond the popular facts and share my personal narrative. My goal now is more than the making of photographs. I want to inspire readers to explore some of Arlington's entombed heroes and learn their stories. In this book, you will occasionally travel outside the cemetery and into the places and homes where the heroes lived. You will move beyond the cemetery and explore a better America. This book is as much about the heroes at Arlington as it is about their families and loved ones; you will find America's better angels. When you walk the lanes of Arlington and move beyond the graves, you experience what the US Constitution calls "a more perfect union." *A Walk Among Heroes* is a portrait of an America worth living in and sadly, at times, dying for.

Arlington National Cemetery is one of the most popular destinations around our nation's capital. Over three million visitors tour Arlington every year. You are confronted with this fact as busloads of tourists unload and mob the visitor center. In the spring and fall, convoys of school buses full of neatly attired students replace

Gravesite of President John F. Kennedy

the tourists. Most children wear uniforms or unique T-shirts to aid their chaperones in herding and minimizing strays.

I've often marched with hundreds of my fellow citizens to the Tomb of the Unknowns or President John F. Kennedy's grave. If you stray from these two sites, however, you are mostly alone among heroes. For several years now, I've walked to the distant graves of the unsung and a few famous patriots. I'm humbled by their sacrifices and grateful they are remembered by a nation that owes them so much. I wish I could learn the stories of all four hundred thousand. Perhaps by remembering a few, I can honor the many.

I used my old Hasselblad camera to take most of the photographs in this book. It's a big, slow, deliberate camera and nothing like today's automatic miracles. Nevertheless, it produces a photograph with silklike qualities. It also creates pictures on film. Using this camera is a slow process—measure the light, determine the exposure, compose, and then compose again. Making a photograph is costly and takes time. The negatives need to be processed. There is no instant feedback. If you mess it up and don't get it quite right, back you go to Arlington to walk among heroes.

Whenever Lillian, my wife, and I go back to our nation's capital, I try to spend some time walking among the graves. I see familiar names, I recall stories of heroes, reminding me that these honored dead were sons and daughters and husbands and wives.

Someday, I, too, will rest at Arlington. But I'm not a hero like these many. I'm not worthy. I'll be there, however. Maybe I can take the midwatch? Maybe those heroes can get some rest, and I'll stand guard. I owe them that. I'll join them at Arlington.

Lillian will join me too. She's always come with me. Why should death be any different? She'll stand watch along with me, standing *by me* as she has always done, not in front nor behind, but together and by my side. And we will be by their sides.

My gravestone at Arlington will have my name, my rank, and a list of a few of my medals. It might also be inscribed with "Vietnam" or "Persian Gulf" for the wars I served through. The other side will simply have a name. Or maybe just "wife of." What else should there be? Mother? Grandmother? What's appropriate for a wife? How about her many academic and professional accomplishments? None will be listed. Will the headstone cover the brilliant children she raised? Often alone? No. Nothing. Just white stone. Her tombstone will simply say "wife of" when so much more should be displayed. She'll be standing watch over the heroes of Arlington, standing watch in her death as she stood watch over me in my life.

The chapters and photographs are my samples of the memorials and gravesites scattered over the cemetery's 639 acres. If you spend a few hours and walk to a distant gravesite, you experience the sheer enormity of Arlington. Over 270,000 headstones fill the landscape, creating what Abraham Lincoln called the "altar of freedom."[1]

Spend a few hours and you will hear "Taps," a gun salute, and maybe see a caisson or two. Almost thirty funerals take place each weekday at Arlington, fewer on Saturday. Before the first funeral of the day, the flag is raised to full staff and then immediately lowered to half-staff. It remains until one half hour after the last funeral of the day. She is then raised again to full staff and then retired. You will not find a personal photograph of a funeral in my collection. It just didn't seem right.

If you live near our nation's capital or plan to visit, I encourage you to tour Arlington National Cemetery. Spend some time walking to distant gravesites and get to know the heroes buried there. Learn their life stories by traveling to their hometowns. Learn how they lived and whom they loved. Find your own better America. If you do, you will see the very best of our nation, where every stone tells a story.

Gravesite of Senator Robert F. Kennedy

HALLOWED GROUND

That we here highly resolve that these dead
shall not have died in vain—that this nation,
under God, shall have a new birth of freedom,
and that government of the people, by the people,
for the people, shall not perish from the earth.

—President Abraham Lincoln

CHAPTER 1

MEMORIAL DAY

Dogwood trees in bloom

WHO STARTED MEMORIAL DAY? IS THERE A SIMPLE EXPLANATION? Was it the South to commemorate its Confederate War dead? Was it the wives and daughters of the Grand Army of the Republic (GAR)? Some say Abraham Lincoln started it with his speech at Gettysburg. But that's just history for convenience. I wanted to use my time on this day to remember the life of one or two heroes buried at Arlington. But their stories would seem out of place without first understanding Memorial Day's larger context—its history and meaning.

At least twenty-five distinct places in the United States lay claim to the origin of Memorial Day. As early as 1861, in Warrenton, Virginia, the grave of John Quincy Marr was decorated with flowers by family and friends on June 3. Marr had died just two days earlier.

Captain Marr was the first Confederate soldier to die in combat. After the Civil War, Southern women and their families cared for the local cemeteries and placed small flowers on the soldiers' graves. Remembering their fallen loved ones, they helped preserve the myth and culture of the lost cause of the Confederacy.

Eight days before that morning, in the North, Colonel Elmer E. Ellsworth led a group of Union soldiers into Alexandria, Virginia, to retake the city. A large Confederate flag was flying atop the Marshall House, an inn in Alexandria, Virginia. On May 24, 1861, Ellsworth and a few of his men entered the building through an open door. They scrambled to the top and removed the flag. Upon descending a flight of stairs, Ellsworth was shot in the chest by the Marshall House innkeeper. He died instantly. In return, one of the Union soldiers shot and killed the innkeeper. Ellsworth was the first Union casualty of the Civil War.

General John A. Logan (by Mathew Brady 1860-1864
Courtesy the Library of Congress)

In 1863, after the Battle of Gettysburg, Lincoln's 272 words would form the ethos of our nation and begin the healing we still hope for today.

On May 5, 1868, General John A. Logan, the commander in chief of the GAR, issued a proclamation for a national day of remembrance called Decoration Day. This was a day to place flowers on the graves of soldiers and loved ones who had died in the war. The GAR was formed after the Civil War to aid northern veterans and advocate for pensions and voting rights for Black Americans. In addition, the GAR urged family members to do what they could to help keep their surviving war veterans sober. A central focus of the GAR was to create a national day of remembrance for the more than 360,000 Union Army dead.

According to the Department of Veterans Affairs, May 5, 1868, is officially recognized as the first Memorial Day. However, the following decades saw numerous towns throughout the North and South laying claim to the honor. For example, Columbus, Georgia, claimed it celebrated Memorial Day in 1866, as did Waterloo, New York.

How did we settle on the last Monday in May for Memorial Day? To commemorate some famous battle or to mark the turning point in the Civil War? No, the end of May was selected because that's when flowers are in bloom throughout the North.

As Civil War veterans grew older and their comrades began to die off, they complained that the younger generation was using the holiday as a time for games, picnics, and revelry. Memorial Day took place at the beginning of summer, after all. In 1923, the GAR and the Indiana state legislature introduced a bill opposing the running of the Indianapolis 500 motor race on Memorial Day. But local officials and the American Legion wanted it to continue. The Indiana governor vetoed the measure, and the race went on. Begun in 1911, the Indy 500 celebrated its hundredth annual race in 2016. The race was suspended only twice, once during World War I and again during World War II.

In 1968 and not leaving well enough alone, Congress passed the Uniform Monday Holiday Act. The act moved four federal holidays, including Memorial Day, to a Monday, thus creating three-day weekends. That, as with most things in history and everything in this story, was subject to controversy. The law took effect in 1971, but it took a few years for all fifty states to comply.

As late as 2002, the Veterans of Foreign Wars opposed holding Memorial Day on a three-day weekend. They believed that "Changing the date merely to create three-day weekends has undermined the very meaning of the day. No doubt, this has contributed a lot to the general public's nonchalant observance of Memorial Day."[1]

On every Memorial Day, flags across the country are raised at sunrise and then lowered to half-staff, where they remain until noon. They are raised again to full staff at noon and stay until sunset. Before every Memorial Day, hundreds of soldiers from the 3rd US Infantry Regiment (also known as the Old Guard) place small flags at the over 270,000 gravestones of Arlington National Cemetery. Flags are placed at each headstone precisely one boot length from the base. Seven thousand more flags are placed at the foot of the Columbarium Courts and Niche Wall. Throughout the United States, volunteers place thousands more flags at the gravesites of our honored dead. They do so at the 172 national cemeteries and many others scattered throughout our nation and overseas.

It's not just Memorial Day's placement at the beginning of summer or its creation of a three-day weekend that makes this holiday abstract. The sheer enormity of our war dead hides our connection. Most of us have no direct or firsthand relationship with the 1.35 million service members who lost their lives defending our country, beginning with the Revolutionary War through today. Over 600,000 Union and Confederate soldiers died in the Civil War. As with most numbers or dates in this chapter, there is plenty of disagreement. An analysis in 2011 using census-based research places the Civil War service member dead at closer to 750,000. And the number of soldiers and sailors killed might have been as high as 850,000.

In World War II, we lost half a million service members. A walk along the Vietnam War Memorial Wall in Washington, DC, contains the names of over 55,000 lives lost. As of January 4, 2024, America is numb after more than three years into a global pandemic. We've lost over 1,163,040 to the novel coronavirus so far. Just like our war dead, the pandemic and the more than one million lost are abstract unless one of those who died was your relative, friend, or colleague.

In 1973, the US Selective Service's authority to induct ended. The lottery, or draft, was suspended in 1976. I turned eighteen in 1974, and like every other eighteen-year-old male, I registered for the draft. My lucky number was 112. Had the war continued, I would've been on the shortlist. Many of my generation alternatively chose Canada over the Southeast Asia rice paddies of President Johnson and President Nixon's War. Who knows what the right choice should've been? Most people I know who had the "right answer" were never faced with that decision. After all, the television and living room couch make patriots of us all. Now, almost fifty years since the draft ended, we seldom know anyone who served in the military, let alone someone who died while on active duty.

According to the casualty status from the US Department of Defense (DOD) as of August 21, 2023, Operation Iraqi Freedom cost us 4,418 service members and another 13 DOD civilians. Operation Enduring Freedom, also known as the war in Afghanistan, all twenty-plus years of it, cost us 2,219 service members, 131 who died in other locations, and 4 DOD civilians.[2]

Today, instead of the draft, we have a system I call "Let someone else's kid go." As fewer and fewer of our nation's youth are needed to serve in uniform, there is an ever-widening divide between the military and civilian communities. Unless you live close to a major military installation, it is quite possible not to know someone actively serving in uniform. Without this personal connection, the meaning of Memorial Day becomes an abstract concept. When the two communities do meet on holidays such as Memorial Day, fireworks and hot dogs add a nice touch.

The flag is flown at half-staff

CHAPTER 2

ARLINGTON HOUSE

Built in the Greek revival style, Arlington House took sixteen years to complete

The Arlington House, once called Custis-Lee Mansion, was built by slave labor in 1802. Sitting on a Virginia bluff overlooking Washington, DC, it was first known as Mount Washington. A plantation of 1,100 acres surrounded the mansion. The main building was built as a shrine to George Washington by Martha Washington's grandson, George Washington Parke Custis, and her first husband, Daniel Parke Custis. It was built in the Greek revival style, popular at the time, with bricks plastered with hardened concrete. Its front entrance has an expansive view of our nation's capital.[1]

George Custis's only surviving child, Mary Anna Randolph Custis, married Lieutenant Robert E. Lee in 1831 and inherited Arlington House in 1857 when George Custis died. Mary and Robert raised seven children in the mansion until the beginning of the Civil War.

When the Lees inherited Arlington from Mary's father, they also inherited nearly two hundred slaves and thousands of acres of land on three plantations. The mansion desperately needed repair, and Colonel Lee took a three-year leave of absence from the Army to restore the buildings and surrounding land. Lee felt it was his charge to revitalize the estate. After George Custis died, the slaves thought they'd be manumitted (freed). However, the written will wouldn't grant them freedom for another five years. Lee used that time to increase the work needed from the slaves dramatically. To control the population and as a form of "behavior control," Lee separated slave families. He hired out many of the plantation's slave children. By 1860, Lee had separated at least one child from every enslaved family, and only thirty-eight slaves remained at Arlington.[2]

Robert E. Lee graduated second in his class from the military academy at West Point in 1829. While military service prevented him from spending much time at Arlington, Mary preferred the mansion to military life. She remained at Arlington until Colonel Robert E. Lee resigned his commission in the United States Army on April 20, 1861. Rather than take up arms against his native state of Virginia, he renounced his oath to the Constitution that he'd sworn as a commissioned officer and took up arms against the United States.

The turning point in Lee's life was resigning his commission and supporting the Confederate cause. Lee professed not to want to draw his sword again except in defense of his home state of Virginia. When General Joe Johnston was wounded in the Peninsula Campaign near Richmond in June 1862, Confederate President Jefferson Davis gave Lee command of the Army of Northern Virginia.

Lee led his men on offensive campaigns into the North not once but twice. First, in 1862, he engaged Union General George B. McClellan's soldiers in the Battle of Antietam. In a single day, more Union and Confederate troops lost their lives than on any other day of the Civil War. McClellan failed to pursue Lee's defeated army as they hobbled back into Virginia. Then again, in 1863, Lee took his army on the offensive and met Union General George Meade at Gettysburg, Pennsylvania.

America lost over fifty thousand sons, fathers, and husbands in three bloody days at Gettysburg. Once more, Lee's defeated Confederate Army of Northern Virginia straggled back across the Potomac. Lee might have started as a reluctant secessionist, but he became an ardent defender of the Southern cause. For Lee, defending the Southern way of life meant defending slavery at all costs.

The war also took a personal toll on Robert E. Lee. His daughter Anne and daughter-in-law died during the war. His son Rooney was captured and held as a prisoner of war. When Lee resigned his commission in 1861, he knew Arlington would be lost. The mansion looked down upon the White House less than three miles away. Secretary of State William Seward told Lincoln that a Confederate cannon shell could easily hit any window. Union soldiers quickly moved to build fortifications on the grounds surrounding the mansion and used the house for shelter.

As the war progressed, so did the casualties. Soon all existing cemeteries around Washington were filled beyond capacity. In May 1864, after the Union Army suffered horrible losses at the Battle of the Wilderness, the Army needed more land to bury their dead. Brigadier General Montgomery C. Meigs's job was to find the required ground. The land selected surrounded Arlington House and sat on the high bluff overlooking Washington, DC. Meigs was promoted to quartermaster general of the Union Army. He loved the idea of using the Arlington property as gravesites for Union soldiers.[3]

Meigs was born in Augusta, Georgia, in May 1816, where his father practiced medicine. However, the family quickly moved back to Philadelphia because his parents abhorred slavery and would no longer live in the South. Young Montgomery excelled at math and science and learned German, French, and Latin. He enrolled at the University of Pennsylvania when he was fifteen years old. In 1832, he was accepted into West Point, where he placed high in math and French. In 1836, he was assigned to the quartermaster's corps after graduating fifth in his class.

His career held little note until 1852 when he was summoned to Washington. His task was to bring fresh water to a city that had grown too large for the existing springs and wells. As a result, the Washington Aqueduct was one of the first in the country, and the

Union Arch Bridge, built to carry water over Cabin John Creek, was the longest brick masonry arch bridge in the world. While supervising the aqueduct's construction, he also oversaw the completion of the dome for the US Capitol.[4]

During the Civil War, Meigs was promoted to general and assigned as the quartermaster general for the Union Army. At six feet two, he was brash, demanding, quick-tempered, and highly opinionated. He craved adulation but also had a reputation for hard work, integrity, and honesty. In 1860, Meigs disagreed with the procurement contracts of Secretary of War John B. Floyd. As punishment, he was sent to build Fort Jefferson in the Dry Tortugas seventy miles past the end of the Florida Keys. After Floyd resigned and joined the Confederacy, Meigs returned to Washington, DC.

During the Civil War, Quartermaster General Meigs's rapid resupply of Union forces was nothing short of miraculous. The armies of Ulysses S. Grant and William Tecumseh Sherman credited him with their ability to pursue the Confederates over hundreds of miles. Grant and Sherman might have won the war, but they won because of the hundreds of miles of railroad track, thousands of wagons and horses, and millions of pounds of food and feed supplied by Montgomery C. Meigs. Quartermaster General Meigs moved mountains, but as the war dragged on, the dead piled up.

Just as Lee started as a reluctant secessionist, Meigs began as a hesitant abolitionist. But the war changed him. When the war started, some of his classmates became generals in the Confederate Army. When they did this, he saw them as traitors. Meigs once admired the colonel known as Robert E. Lee; now, he wanted Lee hanged for treason. Meigs vowed that the Lee family would never return to Arlington. He believed Lee had dishonored the land and the heroes buried there because of his secessionist revolt.

While witnessing the dedication and hard work of the formerly enslaved people digging graves at Arlington, Meigs became a champion of emancipation. He favored not only freedom but education and the full rights of citizenship for former slaves.

Meigs believed that placing Union graves as close to the mansion as possible would make it unthinkable for the Lees to return. However, the soldiers bivouacked in the house objected and forced the burials farther down the north side of the property. Union soldiers had occupied the mansion since 1861, and in 1864, the government formally gained the property from the Lees for failure to pay back taxes. The government required that any taxes owed on property be paid in person for Confederate land now occupied by Union forces. When an aide to Mary Custis Lee attempted to pay the tax, the government refused.

Not convinced the Lees would give up their property, Meigs was determined to bury the dead as close to the mansion as possible. In 1864, he had twenty-six graves placed around the mansion's rose garden perimeter. After the war, Robert E. Lee's brother visited Arlington and believed the estate could still be habitable if the graves were fenced off, so Meigs ordered the placement of as many graves as possible closer to the mansion. In April 1866, he approved the construction of a monument to the Civil War's unidentified soldiers. He had the remains of 2,111 soldiers interred in a vault in Mary Custis Lee's rose garden, joining the 15,000 soldiers and sailors already buried at Arlington.

In the years after the war, the Lees won a lengthy court case to regain their property. The Supreme Court ruled five to four in the Lee family's favor. Lee's eldest son, George Washington Custis

Lee, sold the land back to the government for $150,000, or over $4 million in today's dollars. We can only surmise that Meigs's efforts to inter bodies close to the mansion influenced the Lees to never return to Arlington.

As the president of Washington College, later named Washington and Lee College, Robert E. Lee lived the rest of his life in Lexington, Virginia. He continued to believe in slavery and opposed the education of African Americans. Lee died in 1870.

Meigs went on to build the US Pension Building, where tens of thousands of pension applications were processed for veterans of the Civil War. In addition, this department dispersed tens of millions of dollars to wounded veterans, widows, and orphans. The building eventually became the National Building Museum and a historic landmark.[5]

Today, Arlington National Cemetery sits on a Northern Virginia bluff overlooking our nation's capital. Over 400,000 of America's better angels lie in eternal rest on Arlington's 639 acres.

Also joining the over four hundred thousand is Major General Montgomery Cunningham Meigs. He's interred with his daughter. To visit their grave, head straight from the visitor center on Roosevelt Drive, then take a right on Weeks Drive and a left on Sheridan Drive. You will pass the Tomb of the Civil War Unknowns and Arlington House. After the James Tanner Amphitheater, take Meigs Drive. His grave is on the right in Section 1.

Quartermaster General Montgomery C. Meigs helped win the Civil War and honor its fallen. May God bless him and rest his soul. If you have time, you should also visit the original entrance to Arlington at the McClellan Gate. There, you can see Meigs's name inscribed in gold leaf. Once through this gate, you enter America's most hallowed ground.

McClellan Gate

CHAPTER 3

THE GOOD SON

Robert Todd Lincoln is buried with his wife, the former Mary Eunice Harlan, and their son Abraham II

SURVIVING CHILDHOOD WAS NOT A GIVEN IN THE 1850S AND 1860s. In 1863, Emily Dickinson wrote about the inevitability of death and her calm acceptance of it. There would be no calm acceptance for Mary Todd Lincoln.

> Because I could not stop for Death—
> He kindly stopped for me—
> The Carriage held but just Ourselves—
> And Immortality.

Robert Todd Lincoln was born on August 1, 1843. As he grew up, his father traveled on the judicial circuit, so Robert and his father were never close. Robert was the eldest son of Abraham and Mary Todd Lincoln. Abraham was born into poverty, educated himself, and learned the law. Mary Todd was born into privilege and came from a wealthy Kentucky family who owned slaves. For a time, Abraham's political rival, Stephen Douglas, courted Mary. Abraham and Mary Todd would marry and go on to have four sons, Robert Todd in 1843, Edward Baker in 1846, William Wallace in 1850, and Thomas "Tad" in 1853. Sadly, only one of them, Robert Todd, survived to adulthood.[1]

In 1859, Robert Todd took the Harvard entrance exam and promptly failed fifteen of the sixteen subjects. Undeterred, he enrolled at the Phillips Exeter Academy, a New Hampshire prep school designed to groom young men for the Ivy League. Harvard admitted him the following year.[2]

Robert Todd had not yet turned eighteen when in 1861, the South Carolina militia fired on the Union troops at Fort Sumter.[3] Confederate sons in the South and Union sons in the North rushed to enlist. Robert begged his parents for permission to join the Army.

His father wanted him to postpone his schooling and accept a commission, stating, "Our son is not more dear to us than the sons of other people are to their mothers."[4] But Robert's mother wouldn't hear of it. She didn't want to lose another son, especially to that damn war.

Abraham and Mary Todd had lost their son, Edward Baker, in 1850. Eddie was not yet four when he succumbed to tuberculosis. Then, in 1862, the Lincolns lost William Wallace. Willie died of typhoid fever shortly after his eleventh birthday.[5] At the start of the Civil War, the Lincolns had already lost two of their four sons. Mary Todd couldn't bear the thought of losing another child, so Robert remained at Harvard until the last year of the Civil War.

In January 1865, his mother finally relented, and Lincoln's eldest son, Robert Todd, was commissioned as a captain in the Union Army. In a letter to General Ulysses S. Grant, the president urged Grant to consider Robert for a position on Grant's staff. The posting all but assured Robert Todd would never see combat.

Captain Robert Todd Lincoln was present at Lee's surrender to Grant at Appomattox Court House.[6] At the end of the war, Robert returned to Washington. The covered wagon trip lasted for days and exhausted him. At least that's what he said when he turned down an invitation to sit with his parents at Ford's Theater.

He rushed to his father's bedside when word of his father's shooting reached him. His vigil lasted all night until his father, President Abraham Lincoln, passed away.[7] After his father's assassination, Robert Todd took his mother and his youngest brother, Tad, home to Springfield, Illinois.

Abraham had nicknamed the boy Tad due to his large head and how he was as "wiggly as a tadpole."[8] Tad Lincoln died on

July 15, 1871, before his eighteenth birthday. The cause of death was a combination of tuberculosis and pneumonia. After three of her four sons died, Mary Todd Lincoln became increasingly depressed, despondent, and reckless with money. She bought expensive items but never used them. Mary Todd purchased colorful dresses, even though she wore only black. She took to sewing into her undergarments the $56,000 in government bonds President Lincoln left for her.

Robert Todd, convinced his mother would hurt herself, petitioned the courts to commit her to a psychiatric hospital. After three months, Mary Todd engineered an escape and moved in with her sister. The mother and son's relationship, however, would never be repaired. Mary Todd Lincoln remained with her sister as her health declined. On July 15, 1882, eleven years after the death of her youngest son, she lapsed into a coma and passed away.[9]

Robert graduated from law school at Northwestern University and in 1868, married Mary Eunice Harlan, the daughter of Iowa senator James Harlan. Robert and Mary had three children—two daughters and one son.

After Robert's short stint as the town supervisor for South Chicago, President Rutherford B. Hayes offered to appoint him secretary of state. He turned the position down but later accepted a cabinet position as the secretary of war. He also served as minister to the United Kingdom during the Benjamin Harrison administration.

While in England, Robert and Mary's son, Abraham Lincoln II ("Jack"), died of sepsis at age sixteen. Weeks after Jack's death, Robert wrote to his cousin: "We had a long & most anxious struggle and, at times, had hopes of saving our boy. It would have been done if it had depended only on his own marvelous pluck & patience now that the end has come, there is a great blank in our future lives & an affliction not to be measured."[10]

When Robert's time in England ended, he returned to private practice and, in 1897, became the president of the Pullman Palace Car Company after George Pullman died. In 1911, he became Pullman's board chair and remained in that position until 1922.[11]

While Robert Todd served as secretary of war, he witnessed the assassination of President Garfield at the Baltimore and Washington railroad station on July 2, 1881. At the invitation of another president, he was at the Pan-American Exposition in Buffalo, New York, when President William McKinley was assassinated. Not an eyewitness, Robert Todd was just outside the building.[12]

He would never accept another presidential invitation.

"No, I'm not going, and they'd better not ask me because there is a certain fatality about presidential functions when I am present."[13]

On May 30, 1922, Robert Todd did, however, attend the dedication of his father's memorial in Washington, DC. Chief Justice William H. Taft oversaw the memorial's commissioning and presented it to President Warren G. Harding, who accepted it on behalf of the American people. That event was Robert Todd Lincoln's last public appearance. He was seventy-eight years old.[14]

Robert Todd Lincoln died in his sleep on July 26, 1926, at his home in Vermont. He was interred at Arlington National Cemetery with his wife, Mary, and their son, Jack Lincoln.[15]

Neither fame nor fortune make for a good son. Surviving your mother does.

To visit Robert Todd Lincoln's sarcophagus, take a right out of the visitor center paralleling Eisenhower Drive on your right, which turns into Schley Drive as you pass the Women in Military Service for America Memorial. Take a left at the first sidewalk, Custis Walk. Section 31, S-13 is on the left.

Robert Todd Lincoln's sarcophagus

CHAPTER 4

THE SADDEST PLACE IN AMERICA

Gravesite of Second Lieutenant Christopher Loudon

The war in Afghanistan was the longest in American history. Many service members who died in Afghanistan or Iraq were on their second, third, or fourth deployment. The other sad reality of today's wars is the number of troops that come home badly injured. Medicine has come a long way. Large numbers of service members who, in the past, would have died on the battlefield get patched up and sent home. We've done an excellent job mending their bodies, giving them new limbs, faces, and eyes. We've also done more for those who return home with the invisible signs of war. According to the VA, posttraumatic stress claims the lives of 17.5 veterans every day.

Section 60 contains over nine hundred American Soldiers, Marines, Sailors, and Airmen who died in Iraq or Afghanistan.

This section feels different. It feels much more like the present than the past. Every day, maintenance workers in their ATVs remove piles of yesterday's flowers. The trash bins are full of old flowers from Section 60. You see new photographs and memory stones while walking among the graves.

In Arlington's other sections, families often stop briefly to pay tribute to a service member from past wars. Section 60 holds the fallen of the latest generation of fallen heroes. Families and friends arrive and stay. There's no place else to go. The work of the churches, synagogues, temples, and mosques is now complete. There's always a new funeral in the works. Soon another comrade will join the men and women laid to rest in Section 60.

Loved ones bring lawn chairs, food, an extra beer or two, and anything else they can carry to remind them of their lost hero. They sit, lie down, pray, and talk. You can hear complete conversations. Families show up when Arlington opens and don't leave until the security guards force them out at the end of the day. Section 60 has been called the saddest place in America.

Several years back, we visited during springtime. At the visitor center, you can use a kiosk to search for the name of someone interred. The night before, I'd done my homework. Second Lieutenant Christopher E. Loudon died in Iraq in 2006. He'd earned a Bronze Star and a Purple Heart. I didn't know him, but I wanted to see where in Section 60 he is buried—the place of so much pain.

You can get a car pass to visit a particular grave in Arlington. The number of passes handed out each day is limited. The day we visited, we were fortunate to get one. We drove through the cemetery about as slow as a car can go. It still felt too fast. As we drove, "Silence and Respect" signs lined our route.

I forgot to tell Lillian I wasn't looking for Lieutenant Loudon. He was my excuse to get a pass and drive to Section 60. I wanted to see all the graves, witness the sadness, and walk among our fallen from the wars in Iraq and Afghanistan.

We parked, and Lillian took the directions printed from the kiosk and began searching for the young lieutenant. As she followed the numbers looking for Chris's grave, I headed into Section 60's sea of brand-new headstones. Other rows bore only the plastic markers of new graves. A permanent headstone would be more than a year away. The workload is backed up in Section 60. The burials and headstones will come but not soon enough.

As we walked about, a dozen or so families paid their respects. A young mother held her two-year-old daughter. An older couple wandered, looking for their great-grandson. A Vietnam vet had camped at one of the graves and smoked a cigar. It wasn't his first visit, nor would it be his last.

Once Lillian realized I wasn't looking for Lieutenant Loudon's grave, she walked over to where I stood.

"I thought we came to see Lieutenant Loudon?"

"I wanted to see them all."

"You could have told me . . ."

"Sorry."

"He's over there."

"Look at this grave. Look how many stones."

The Jewish have a custom of placing small stones on headstones. They are an enduring symbol of how the memory of someone will last. Flowers wither and fade, as does life.

An Asian couple set up chairs for a picnic lunch. They'd come to spend the day. It turned out they came every Saturday. They hadn't missed a Saturday for over a year now.

"Are you from around here?" I asked.

"We live in Alexandria."

"We do too. Whereabouts?"

"Right off Duke Street. How about you?"

"Further south. In Hayfield. Our son went to high school at Hayfield High. How about your son?"

"Edison."

"Did your son play football?"

"Why yes, he did. Our son lost his homecoming game to Hayfield."

"Sorry about that. Do you mind? Can I sit? Would you tell me about your son?"

"Would you like a beer?"

"I'd love one."

Sadly, I can't remember the young soldier's name or the names of his mother or father. But I can remember the kind of beer we drank. So, there I sat. I made polite conversation with the mother and father, who'd lost their only son. Then, I sipped my Corona and enjoyed the mild spring weather.

Section 60 is about three-quarters of a mile from the visitor center. I recommend you walk if you're not visiting a loved one or friend. Take your first left on Eisenhower Drive and then a right on Sergeant Alvin York Drive. Lieutenant Loudon's grave is in the middle of Section 60, surrounded by his friends and comrades.

New graves in section 60

GALLANTRY

A high station in life is earned by the gallantry with which appalling experiences are survived with grace.

—Tennessee Williams

CHAPTER 5

THE MEDAL OF HONOR

Medals of Honor of the Army (*top left*), Navy (*top right*), and Air Force (*bottom*)

WITH THE PASSING OF HERSHEL "WOODY" WILLIAMS, ONLY sixty-three Medal of Honor recipients are still living. Chief Warrant Officer Williams passed away on June 29, 2022. He was ninety-eight and the last surviving WWII Medal of Honor recipient. Born to a dairy farmer in West Virginia, Woody was only three and a half pounds at birth and was the youngest of eleven children. His mother didn't think he'd survive. She named him after the doctor, who showed up several days after his birth. By the time Woody turned eleven, his father had died of a heart attack.

In 1942, Woody was rejected by the Marines for being too short. He stood five feet six. He favored the Marine dress blue uniforms and thought the Army's brown wool "the ugliest thing in town." In 1943, the Marines changed their height standards, and Woody enlisted.[1]

The Medal of Honor is awarded for gallantry and intrepidity. It is the highest honor our nation can bestow. The Medal of Honor citation has a certain rhythm and cadence to it. For example, here is Hershel "Woody" Williams's Medal of Honor citation:

The President of the United States in the name of The Congress takes pleasure in presenting the

MEDAL OF HONOR

to

CORPORAL HERSHEL W. WILLIAMS

UNITED STATES MARINE CORPS RESERVE

for service as set forth in the following

CITATION:

For conspicuous gallantry and intrepidity at the risk of his life above and beyond the call of duty as Demolition Sergeant serving with the First Battalion, Twenty-First Marines, Third Marine Division, in action against enemy Japanese forces on Iwo Jima, Volcano Island, 23 February 1945. Quick to volunteer his services when our tanks were maneuvering vainly to open a lane for the infantry through the network of reinforced concrete pillboxes, buried mines and black volcanic sands, Corporal Williams daringly went forward alone to attempt the reduction of devastating machine-gun fire from the unyielding positions. Covered only by four riflemen, he fought desperately for four hours under terrific enemy small-arms fire and repeatedly returned to his own lines to prepare demolition charges and obtain serviced flame throwers, struggling back frequently to the rear of hostile emplacements, to wipe out one position after another. On one occasion he daringly mounted a pillbox to insert the nozzle of his flame thrower through the air vent, kill the occupants and silence the gun; on another he grimly charged enemy riflemen who attempted to stop him with bayonets and destroyed them with a burst of flame from his weapon. His unyielding determination and extraordinary heroism in the face of ruthless enemy resistance were directly instrumental in neutralizing one of the most fanatically defended Japanese strong points encountered by his regiment and aided in enabling his company to reach its objective. Corporal Williams' aggressive fighting spirit and valiant devotion to duty throughout this fiercely contested action sustain and enhance the highest traditions of the United States Naval Service.[2]

The Medal of Honor is authorized for any military service member who "distinguishes himself conspicuously by gallantry and intrepidity at the risk of his life above and beyond the call of duty

1. while engaged in an action against an enemy of the United States;
2. while engaged in military operations involving conflict with an opposing foreign force; or
3. while serving with friendly foreign forces engaged in an armed conflict against an opposing armed force in which the United States is not a belligerent party."[3]

To understand what sets the Medal of Honor apart from other military medals, it is important to understand a bit of its history and the medal's enduring significance. All military medals awarded to individuals fall into two broad categories. The first is for meritorious service. Usually, these medals are for a cumulation of superior effort over a significant amount of time. Medals such as the Army or Navy commendation medal and the meritorious service medals are the most prevalent. The other group of medals are meant to recognize valor, not superior service. During the Civil War, Congress felt that it was important to recognize individual acts of bravery.

The Medal of Honor was first created during the Civil War and used much more widely than it is today. Nearly 40 percent of Medal of Honor recipients were for actions during the Civil War. Originally, the medal was intended to recognize the heroic acts of enlisted men only. It was not until 1915 that the rules were changed to include Navy and Marine Corps officers, and the Army followed suit shortly after.

Between 1916 and 1917, the Army reviewed every Medal of Honor awarded and removed 911 recipients, determining that they were erroneously bestowed. In 1918, Congress authorized the Distinguished Service Cross, the Navy Cross, and the Silver and Bronze Stars for valor.

Since then, the Medal of Honor has been solely and distinctly reserved for service members whose actions, at significant risk to their own lives, exemplified the highest level of intrepidity and gallantry. While gallantry conjures up visions of King Author's court and the Knights of the Round Table, I had to look up the word *intrepidity*, one that I and most other people don't commonly use. Intrepidity as defined in the *Merriam-Webster Dictionary* means "characterized by resolute fearlessness, fortitude, and endurance." If you are looking for a synonym, there's *bold*, *brave*, *courageous*, *fearless*, *stalwart*, *undaunted*, or *valiant*. I rarely use any of these words in day-to-day conversations. They just don't come up. If you visit Arlington, one of the most uncommon things is how common these words are. Four hundred Medal of Honor recipients rest in the hallowed ground of Arlington. Four hundred out of the four hundred thousand.

It is important to note that the Medal of Honor is not won. It is not a prize. If you refer to it as such, it's considered an insult. Medal of Honor recipients often explain that their medal is more for their fallen comrades than themselves. The president of the United States awards it on behalf of Congress. There is no such thing as the congressional Medal of Honor, it is simply called the Medal of Honor.

As of 2023, Medal of Honor recipients include 2,461 from the Army, 749 from the Navy, 300 from the Marines, 19 from the Air Force, and 1 from the Coast Guard.[4]

The Medal of Honor is always worn above a service member's highest medal. The design has remained essentially unchanged for over 150 years. Of the millions of men and women who have ever served in the United States military, only 3,530 were awarded the Medal of Honor. Only 126 medals were awarded during WWI, and less than 500 were awarded in WWII. During the nineteen years, five months, four weeks, and one day of the Vietnam War, 266 Medals of Honor were awarded.[5]

The Medal of Honor is the oldest combat decoration of the United States Armed Forces. It was first proposed in 1861 by Iowa Senator James W. Grimes, who, as the chairman of naval affairs, wanted a medal "to be bestowed upon such petty officers, seamen, landsmen, and marines as shall most distinguish themselves by their gallantry in action and other seaman-like qualities during the present war." Two months and nine days later, Senator Henry Wilson, the chair of the Senate Committee on Military Affairs and the Militia, introduced a resolution for a Medal of Honor for Army soldiers.[6]

The first actions to be recognized for gallantry qualifying for the Medal of Honor occurred before the medal was created. In February 1861, Lieutenant Bernard J. D. Irwin led a voluntary platoon to rescue sixty soldiers trapped in Apache Pass, Arizona. The award still needed to be proposed and signed into law by President Lincoln. It would be thirty years before his medal would be presented.

The original intent of the Medal of Honor was solely for enlisted troops, but acts of valor often defy rank, race, and religion. The following are a few notable recipients of the Medal of Honor:

First recipient: Private Jacob Parrott, Civil War—Parrott was one of nineteen men who went two hundred miles into Confederate territory to capture a railroad train and attempted to destroy bridges and tracks at Big Shanty, Georgia.

First and only woman recipient: Contract Surgeon Mary E. Walker, Civil War—From 1861 to 1864, Walker devoted herself tirelessly to caring for wounded soldiers on the battlefield and in the hospital. For four months, she was a prisoner of war. Because of her gender, the law would not allow her a commission as an officer. President Andrew Johnson awarded her the Medal of Honor instead.

First African American recipient: Sergeant William H. Carney, Civil War—In the heat of battle at Fort Wagner, South Carolina, when the Color Sergeant of the 54th Massachusetts infantry was shot, he grasped the flag and made his way forward. Then, under intense fire and wounded twice, he planted his regiment's standard on a small hill overlooking the battlefield and rallied his comrades.

First Hispanic or Latino recipient: Corporal Joseph E. DeCastro, Civil War—As the standard-bearer for Company 1 of the 19th Massachusetts Infantry, DeCastro was on the line during the last day of the Battle of Gettysburg. During Pickett's Charge, he used his own flag to attack the standard-bearer of the 14th Virginia Infantry, capturing their flag.

First Jewish recipient: Private Benjamin B. Levy, Civil War—On June 30, 1862, Levy was the drummer boy for the 1st New

York Infantry. When all his unit's color-bearers were shot, he rushed forward into the fight, saving his unit's colors and preventing the enemy from capturing their flag.

First Asian Pacific recipient: Seaman James Smith, 1872—While serving on board the USS *Kansas*, Smith saved the lives of his fellow seamen and officers as the ship sank.

First Native American Indian recipient: Sergeant Co-Rux-Te-Chod-Ish (Mad Bear), 1869—Co-Rux-Te-Chod-Ish charged forward from his line toward a dismounted Indian combatant. He was shot and gravely wounded by a bullet from his own unit.

Youngest recipient: Drummer Boy William "Willie" Johnston, Civil War—Johnston served in the 3rd Vermont Infantry during the Seven Days Battles of the Peninsula Campaign. He held and secured his instrument as others in his company turned and ran. He was thirteen years old.

First father and son recipients: Lieutenant Arthur MacArthur Jr., Civil War, and General Douglas MacArthur, WWII—Arthur MacArthur distinguished himself when he was only eighteen years old. During the Battle of Missionary Ridge, he seized his regiment's colors and ran forward, planting them on the crest. His son Douglas MacArthur mobilized, trained, and led defensive and offensive operations against a vastly superior Japanese force.

First president and son recipients: Lieutenant Colonel Teddy Roosevelt, Spanish-American War, and Brigadier General

> Theodore Roosevelt Jr., WWII—Teddy Roosevelt led his dismounted cavalry troops up San Juan Hill, rallying them through open fields and enemy fire. He was the first to reach the enemy trenches and quickly eliminated the enemy. Theodore Roosevelt Jr. led the first wave of soldiers storming Utah Beach at Normandy. Under constant fire from fortified enemy positions, he moved throughout the beach, directing and rallying his men. Under his leadership, his men destroyed enemy strongholds and rapidly moved his men off the beach.[7]

The Tomb of the Unknowns at Arlington Cemetery contains the remains of unidentified US service members from WWI, WWII, the Korean War, and the Vietnam War. These unidentified service members are awarded the Medal of Honor. Also awarded the United States Medal of Honor are unidentified WWI comrades in arms from France, the United Kingdom, Italy, and Romania, entombed elsewhere.

The Tomb of the Unknowns combines an aboveground tomb or sarcophagus and three in-ground crypts. The tomb contains the remains of an unknown soldier from WWI. The right crypt contains the remains of an unknown soldier from WWII. The left crypt contains the remains of an unknown service member from the Korean War. The center crypt remains empty after First Lieutenant Michael Blassie was identified and reinterred at Jefferson Barracks National Cemetery in Saint Louis, Missouri. The center crypt was rededicated to all unidentified service members who died in the Vietnam War.

As of 2023, Medal of Honor recipients receive a monthly tax-exempt pension of $1,619.34. They are also entitled to burial at Arlington National Cemetery, and their children are automatically granted admission to one of the service academies.

I might not be able to visit all the Medal of Honor graves at Arlington, but I can learn their stories.[8]

I encourage you to discover more about the Medal of Honor recipients yourself. A good resource on the Medal of Honor, its meaning, its history, and the recipients is *The Medal of Honor: 160 Years of Courage and Sacrifice* by the Congressional Medal of Honor Society (CMOHS), a nonprofit dedicated to preserving the legacy of the Medal of Honor.[9] For more information about CMOHS, visit www.cmohs.org.

To visit the Tomb of the Unknowns, walk straight from the visitor center on Roosevelt Drive. The tomb is a little over half a mile down and next to the Memorial Amphitheater.

Tomb of the Unknowns

CHAPTER 6

JIMMY DOOLITTLE

Doolittle Raid Display at the National Museum of the US Air Force, Dayton, Ohio

While the headquarters for the National Reconnaissance Office (NRO) was in Washington, DC, the Air Force's NRO component was in Los Angeles. During the early 1980s, I was stationed at the Strategic Air Command headquarters in Omaha, Nebraska. NRO officers I worked with flew out of Los Angeles and headed to somewhere else nearly every week—a design review here or an interface meeting there. We always had some meeting across the country. For example, Raytheon was in Garland, Texas, and Boeing was in Seattle, Washington. I would meet my NRO colleagues someplace far from any of our homes. My favorite trips were the ones to Harris Corporation in Melbourne, Florida. I could leave the cold Nebraska winters and walk on the beach. Those trips were fun—at first.

If you were in the Air Force and worked for the NRO, you took the redeye. Monday morning flights from Los Angeles International Airport were the norm. But the big decision was when to come home. Your meetings were over Friday afternoon, so you had a choice. Waiting to fly home on Saturday morning meant losing half your weekend, so most of us opted for the Friday night flight, which arrived back home early Saturday morning. Flying all night gave you an entire weekend at home, in theory, if you could stay awake.

For many years, I matched travel miles with all my program office comrades. We'd meet at Chicago's O'Hare on a Monday morning and travel around the country. I would fly out on a Monday and return early Saturday morning, which led to too many weeks spent away from home—too many weekends lost recovering from the redeye.

When in Los Angeles, I'd passed General Jimmy Doolittle's uniform on display at the entrance to the AF/NRO program offices in El Segundo, California. I didn't know much about him then. I only vaguely knew of his daring raid and little else.

Josephine Elsie Daniels, called Joe by her father, married Private First Class Jimmy Doolittle on December 24, 1917. They were high school sweethearts. Together, and for over seven decades, they were America's first family of aviation. He became a general. She was always Mama Joe.

What did Joe Daniels see in young Private Jimmy Doolittle that would stand the test of time and endure so much hardship? If Jimmy Doolittle was the embodiment of an American aviator, Mama Joe was the heart of America's aviation home. At the Doolittle's, they always had an extra place at the dinner table or a spare bed for a stray pilot.

General James Harold Doolittle was born in Alameda, California, on December 14, 1896, but spent his early years in Nome, Alaska. He returned to California to finish high school and attend Los Angeles City College. Jimmy Doolittle earned admission to the University of California, Berkeley, but took a leave of absence in 1917 to enlist in the Army's Signal Corps Reserves and get married. In 1918, he earned his commission as a second lieutenant and received his aviator wings.

He spent World War I stateside, training pilots. Then, in the early twenties, he returned to Berkeley and finished his bachelor of arts degree. In 1922, he was the first to make a transcontinental flight across America. He made one stop to refuel at Kelly Airfield in San Antonio, Texas. On his flight from Jacksonville Beach, Florida, to San Diego, California, he completed the coast-to-coast flight in twenty-one hours and nineteen minutes. For his record-breaking flight, he was awarded the first of three Distinguished Flying Crosses.[1]

In 1929, the stock market crashed, kicking off the Great Depression. But 1929 was also the birth of modern aviation. Pilots could now take off and land and fly at night, in any kind of weather, with zero visibility. These feats are common today but were unheard of in 1929. Jimmy Doolittle was the first to pilot a plane solely by instruments. "Bold and fearless," they called him. I would not have used exactly those words, but somebody had to be the first. Jimmy Doolittle made all my redeye trips home possible.

To celebrate the occasion of the first flight solely by instruments, Mama Joe held a dinner party. After dinner, she asked her guests to autograph her white linen tablecloth. She kept up this tradition for six decades.

When the Japanese attacked Pearl Harbor, Jimmy Doolittle was a lieutenant colonel. He'd also earned a doctorate in aeronautics from the Massachusetts Institute of Technology. With the destruction of Pearl Harbor and the surrounding airfields, the Japanese owned the Pacific. While the US Navy quietly regrouped and secured their remaining fleet, America feared attacks along our western coast, from San Diego to Alaska. The Japanese leadership and military told their people that their islands were impenetrable. But the Japanese high command forgot to tell Lieutenant Colonel James H. Doolittle. In less than four and a half months, in just 133 days, he proved them wrong.

Eighty men volunteered for one of the most secretive missions this country has ever created. The operation involved a total of sixteen B-25 aircraft. Each plane held a five-man crew, and all were volunteers. The mission was so secret the volunteers couldn't know any details. History will never forget these eighty aviators known as Doolittle's Raiders.

At Eglin Field, Florida, the B-25 crews practiced taking off in five hundred feet or less. Revving the twin engines to full throttle and then releasing the brakes was more like a cannon shot than anything under a pilot's control. While takeoffs were courtesy of the US Navy catapult officers, better known as *shooters*, the planes' landings would be all on the volunteers—no need to train for that.

Days after Pearl Harbor, General Hap Arnold, commander of all the US Army Air Forces, approved Lieutenant Colonel Doolittle's plan. They'd use B-25s to take off from the aircraft carrier USS *Hornet*'s deck and strike the heart of Japan. And strike they did.

Doolittle's men watched as their sixteen planes were lowered onto the deck of the USS *Hornet*. The Raiders didn't realize the importance of their mission until the USS *Hornet* met up with their escort, the aircraft carrier USS *Enterprise*, and her battle group. Their perilous mission was to take off four hundred miles from Tokyo and strike the capital and other primary industrial targets in the heart of Japan. Then, the Raiders would fly on to airfields on the Chinese mainland.

The planes were stripped down in weight to survive the long flight. They removed the bottom gun turret and replaced the rear guns with broomstick handles. To avoid falling into the hands of the Japanese, gone were the top-secret Norden bombsights. Instead, they installed a makeshift aiming sight called the "Mark Twain." Wired to five of the bombs were Japanese friendship medals presented to US servicemen before the war.[2]

All was going according to plan until the carrier battle group was about 650 miles from Japan. A Japanese patrol boat spotted the ships. With their cover blown, Doolittle decided to launch his fleet of B-25s early. His decision meant that the planes would fly an extra 250 miles and not have enough fuel to reach their landing areas. After the bombing raid, they would have only two options: bail out or crash. Their chances of being captured or killed were great. All eighty Raiders knew the risk.

They wouldn't have any fighter escorts. Over half the bombers' defensive capabilities were removed. Because of the distance and the need for extra fuel, each bomber carried only four five-hundred-pound bombs. Sixteen crews of five men delivered thirty-two thousand pounds of bombs. The mission was not an overwhelming counterattack but rather a sharp poke in the bloody red eye of Japan's Rising Sun. America needed a morale boost, and the Japanese people needed to know their island was vulnerable even though their military told them otherwise.

The attack caused the Japanese military to recall a few of their carriers to defend their homeland, which reduced the Japanese threat to our western seaboard. The raid, unfortunately, accelerated the Japanese navy's plans to attack Midway Island and also contributed to one of American history's most significant naval victories. At Midway, the Japanese lost four of their six carriers used to attack Pearl Harbor. But that's a different story.

The Doolittle raid came at a high price. Fifteen B-25s on the mission crashed on the Chinese mainland or in the South China Sea. One plane landed in the Soviet Union and was returned with its crew a year later. One crew member died when his B-25 crashed. The Japanese took eight Raiders as prisoners and executed three of them.

Jimmy Doolittle was sure he'd be court-martialed after he crashed and learned the fate of most of the other crews. Instead, President Roosevelt awarded him the Medal of Honor. The president also promoted Jimmy Doolittle two ranks to brigadier general. He skipped the rank of colonel altogether. All crew members received the Air Medal. Back home, they were heroes.[3]

Medal of Honor Citation
Rank and organization: Brigadier General,
US Army Air Corps
Place and date: Over Japan
Entered service at Berkeley, Calif.

Birth: Alameda, Calif.
GO No: 29, June 9, 1942
For conspicuous leadership above the call of duty, involving personal valor and intrepidity at an extreme hazard to life. With the apparent certainty of being forced to land in enemy territory or to perish at sea, Gen. Doolittle personally led a squadron of Army bombers, manned by volunteer crews, in a highly destructive raid on the Japanese mainland.

China also paid a terrible cost in the raid. For their part in aiding the American Raiders, the Japanese slaughtered over seventy thousand Chinese soldiers and 250,000 Chinese civilians.

Brigadier General Doolittle was later promoted to lieutenant general and went on to command the 12th Air Force in North Africa, the 15th Air Force in the Mediterranean, and then the 8th Air Force in Europe. Finally, in 1983, long after his retirement and forty-three years after the Tokyo raid, President Reagan promoted him to the rank of full general, four stars. General James H. Doolittle died in 1993 at the age of ninety-six.

In 1946, in celebration of Jimmy Doolittle's fiftieth birthday, the sixty-one surviving Raiders gathered—for what would become an annual tradition. The City of Tucson, Arizona, presented each surviving Raider with a silver goblet with his name etched twice—once right side up, then again upside down. Upon the death of one of the airmen, the goblet remains upside down. The Raiders sip Hennessy V.S cognac, vintage 1896, for the year of Jimmy Doolittle's birth. After the toast, they turn over the goblet in honor of a Raider who died. In 2013, the four remaining Raiders gave their

Eighty silver goblets on display at the
National Museum of the US Air Force, Dayton, Ohio

last toast. The goblets are now on display at the National Museum of the US Air Force in Dayton, Ohio.

We take much for granted these days. A coast-to-coast flight or flying at night are now all too common. Also taken for granted are the sacrifices of an older generation. Sometimes, the best way to honor the bravery of heroes such as Jimmy Doolittle or the warmth and comfort of Mama Joe is to remember their stories.

On Christmas Eve 1988, Josephine Daniels Doolittle passed away. It was her seventy-first wedding anniversary.

General James Harold and Josephine Elsie (Daniels) Doolittle rest in Section 7A, just below the Tomb of the Unknowns—a little over half a mile from the visitor center. A single common headstone marks their grave, with General Doolittle on the front and his wife, Josephine Daniels Doolittle, on the back.

We can't all be brave like Jimmy Doolittle, but we can all be kind. Mama Joe would like that.

Gravesite of General James H. Doolittle

CHAPTER 7

TIP OF THE SPEAR

Gravesite of Captain Johnny Micheal Spann.
He is also the 79th star on the CIA's Wall of Honor

It took only a few hours to know who attacked us on September 11, 2001. On October 7, 2001, less than a month later, the United States launched Operation Enduring Freedom. President Bush held up the invasion of Afghanistan to give the Taliban one last chance to surrender the terrorist Osama bin Laden. The president did not delay the deployment of the Central Intelligence Agency's Special Activities Division. This elite group of operatives is often the first to the fight. Together with their comrades in the United States Special Operations Command, they are the tip of the spear. Johnny Micheal Spann was one of the first pairs of American boots on the ground in Afghanistan.

Spann joined the Marine Corps Reserves in 1991 while attending Auburn University. After graduating, he was sent to Officers Candidate School and was commissioned a second lieutenant. He served six years in the Marines, achieving the rank of captain. In 1999, Spann joined the CIA and trained at "The Farm." The Farm is where officers of the US Clandestine Service, as well as other special agents, train.

In October 2001, two Army Black Hawk helicopters transported Spann and a small unit of interrogators and other CIA officers south from Uzbekistan into northern Afghanistan. Their job was to gather intelligence and prepare for the arrival of more US troops. These eight CIA officers were the first Americans behind enemy lines after September 11. Additional CIA clandestine officers and Army Green Berets members arrived shortly after Spann's team. On November 25, 2001, Johnny Micheal Spann was killed during a revolt of several of the over four hundred al-Qaeda terrorists who had surrendered the night before. He was the first American killed in our twenty-year war in Afghanistan.[1]

For his actions during the Battle of Qala-i-Jangi, Spann was awarded the Intelligence Star and the Exceptional Service Medallion. The qualifications for the Intelligence Star read, "Voluntary acts of courage performed under hazardous conditions or for outstanding achievements or services rendered with distinction under conditions of grave risk." The Intelligence Star is the equivalent of the US Armed Forces Silver Star, thus qualifying Spann to be buried at Arlington.

Joining Spann at Arlington are a few other CIA officers and Intelligence Star recipients. These include William Francis Buckley, who was a CIA operative collecting intelligence on the terrorist group Hezbollah. He died around June 3, 1985, after being captured and tortured by the Islamic Jihad, a Shia militia active in the 1980s during the Lebanese Civil War.

Also interred at Arlington is Francis Gary Powers, an Air Force pilot who flew a U-2 reconnaissance plane for the CIA. The Soviets shot down his top-secret spy plane. Powers was captured, convicted of espionage, and held as a prisoner of war until he was released in a hostage exchange for a Soviet colonel in the KGB. Powers died in 1977 in a helicopter accident as a pilot working for the Los Angeles television news station KNBC. In addition to Powers's Intelligence Star, the CIA awarded him its Director's Award. His other military medals include the Air Force Silver Star, the Distinguished Flying Cross, the National Defense Service Medal, and the Prisoner of War Medal.

Arlington National Cemetery's burial requirements are the most restrictive of all veteran and national cemeteries. While all US service members are entitled to burial at a national cemetery, only Armed Forces veterans awarded the Medal of Honor,

Distinguished Service Cross, Distinguished Service Medal, Silver Star, or Purple Heart may be interred (in ground) at Arlington. Military members killed in action, US service member retirees, their spouses, and minor-age children are also entitled to an in-ground burial at Arlington.

In addition to being awarded the Intelligence Star for bravery under extreme risk of life, Johnny Micheal Spann and William Francis Buckley each have a single star carved into the granite wall next to their names on the Memorial Wall at CIA headquarters in Langley, Virginia. To be awarded this honor, a CIA member must have died under the following conditions:

> Death may occur in the foreign field or in the United States. Death must be of an inspirational or heroic character while in the performance of duty; or as the result of an act of terrorism while in the performance of duty; or as an act of premeditated violence targeted against an employee, motivated solely by that employee's Agency affiliation; or in the performance of duty while serving in areas of hostilities or other exceptionally hazardous conditions where the death is a direct result of such hostilities or hazards.[2]

Of the 140 individuals honored on the Memorial Wall, those from the CIA who lost their lives in service to our country, only 103 names are publicly known. Thirty-seven remain secret. A simple two-and-a-quarter-inch star memorializes each of them.

Working in secret, the Central Intelligence Agency cultivates a culture of quiet success. Those successes began with one man, William Joseph Donovan.

Donovan was born in Buffalo, New York, on January 1, 1883. Of Irish descent, he attended a Catholic college preparatory school where he acted in plays, excelled in football, and won awards for his oratory skills. He originally considered becoming a priest but decided "he wasn't good enough to be a priest."[3] After graduating from Niagara University with a prelaw degree, he went on to Columbia Law School. In 1909, as a newly minted lawyer, he returned to Buffalo and joined the law firm of Love & Keating. Two years later, he opened his own firm with a fellow Columbia University classmate.

In 1912, Donovan helped form and lead a cavalry unit of the New York National Guard. In 1914, he married Ruth Rumsey, a Buffalo heiress. His Guard unit was mobilized in 1916, and he took his cavalry troop to the Texas border to join Brigadier General John J. Pershing's army in the hunt for Pancho Villa.[4]

During World War I, Donovan, now a major, led the 1st Battalion, 165th Infantry of the 42nd Division. He proved himself a fearless and courageous leader. He suffered a shrapnel wound in one leg and was almost blinded by mustard gas. He was offered the French Croix de Guerre but turned it down until a fellow Soldier who was Jewish was awarded it as well. He was awarded the Distinguished Service Cross for leading an assault during the Aisne-Marne campaign. His reputation earned him the nickname "Wild Bill," something he had mixed feelings about.

After the war, Donovan returned to Buffalo to resume his law practice; he also served as US Attorney for the Western District of New York. In 1923, he was awarded the Medal of Honor for his heroic actions in the battle of Landres-et-Saint-Georges. The ceremony was attended by more than four thousand veterans. He refused to keep the medal, saying that it belonged "not to me but the boys who are resting under the white crosses in France or in the cemeteries of New York, and those lucky enough to come home."[5]

In the lead-up to World War II, Donovan developed a keen understanding of world affairs and collected intelligence on the Fascist dictator Mussolini and the growing Nazi influence. President Roosevelt admired Donovan and used him as a trusted adviser. The two disagreed on domestic policy and were from opposing political parties, but their personalities were similar, and their mutual respect maintained their close relationship throughout the war.

Donovan was now a colonel and began to grow his network of intelligence-gathering spies. President Roosevelt felt that American intelligence in the lead-up to and beginning of WWII was uncoordinated and ineffective. He tapped Donovan to head up the Office of Strategic Services (OSS), reporting directly to the Joint Chiefs of Staff. In June 1942, the OSS was formed with Colonel William Donovan named as the Coordinator of Information. Donovan built a centralized organization to gather intelligence and conduct special operations. He recruited commandos to drop behind enemy lines to provide disinformation and disrupt enemy operations. He cultivated partisans and funneled money and supplies for guerrilla activities.[6]

Much of the organization and structure of the OSS is modeled after the British spy services. Both Donovan and Roosevelt relied heavily on British spymasters. Colonel Charles Howard "Dick" Ellis was an Australian-born intelligence officer of the British Secret Intelligence Service and is credited with the blueprint for US intelligence organization and operations. The OSS conducted some of the most risky and costly operations of WWII. After the sudden death of Roosevelt, Harry S. Truman became president. He distrusted Donovan despite the praise and admiration of General Eisenhower. The OSS was also viewed as a direct threat to the power and authority of J. Edgar Hoover and the FBI. After the war, because of its power and potential for overreach, the OSS was disbanded.

In 1947, facing a new Soviet threat, the National Security Act was signed into law. This law reorganized our country's military infrastructure. The War Department and the Navy Department were merged into the Department of Defense. The National Security Council was also created to provide coordinated intelligence directly to the president of the United States. This act made the United States Air Force a separate service branch and no longer part of the US Army. With this new law came a recognition that the vision of William Donovan was correct. The National Security Act of 1947 created the Central Intelligence Agency to understand and, if need be, counter foreign threats.

Long after I retired from the United States Air Force with over twenty-two years on active duty, I joined the Office of Strategic Services Society. The society celebrates the historic accomplishments of the OSS during World War II and is a charitable nonprofit organization. You need to be a direct descendant of a member of the OSS or you must provide proof of your service as either an intelligence officer or your work in special operations. To be a member of the OSS Society, you must offer proof of your service at the tip of the spear.

History is changed by men and women of vision. Donovan provided inspiration and foresight when our nation needed it the most. At CIA headquarters stands a statue dedicated to William J. Donovan, the founder of the CIA.

To visit Major General William J. Donovan's grave is a seven-minute walk. Head straight from the visitor center on Roosevelt Drive and turn right on Grant Drive. His grave is on the left in Section 2.

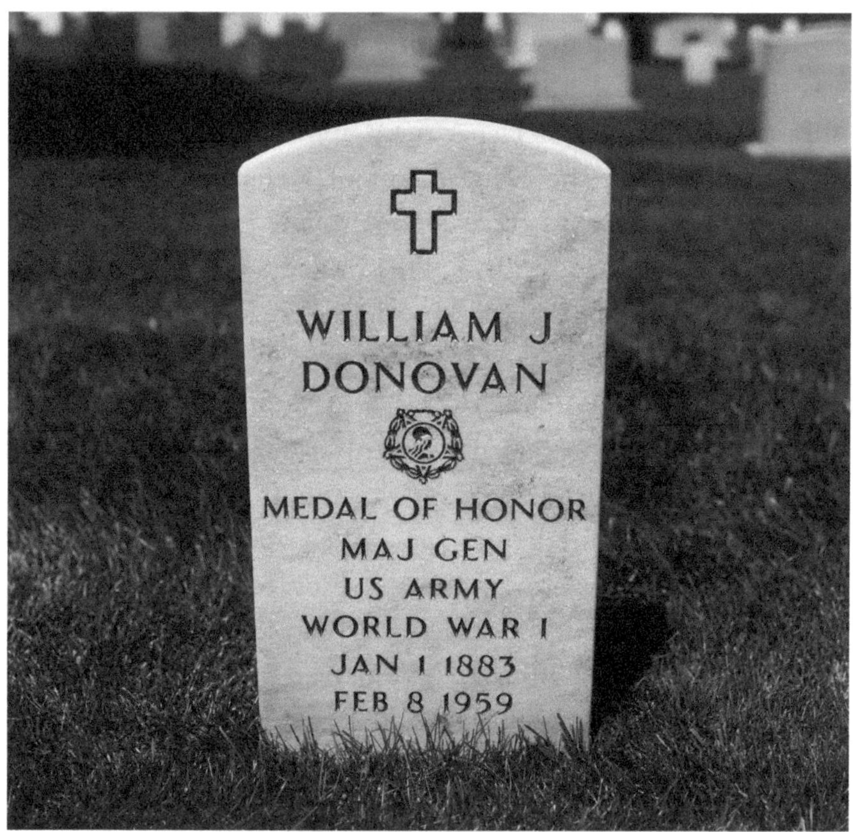

Gravesite of Major General William J. Donovan

CHAPTER 8

TWENTY SECONDS

Gravesite of Sergeant John L. Levitow

Twenty seconds, even less—that's all the crew of *Spooky 71* had. The gunship was coming back around. They'd already eliminated two North Vietnamese mortar batteries, and now a third fired into the Long Binh Post, a major US Army base outside of Saigon. In 1969, the attacks were more a death from a thousand cuts than the overwhelming sweep of the Tet Offensive the year before.

By 1969, the base was under constant attack. The Army doubled its border fences and sent out night patrols. Long Binh Post was the major logistics base in South Vietnam. The plan was to get the thousands of soldiers out of the city of Saigon and house them all twenty miles to the northeast at Long Binh. Knowing this, the North Vietnamese and Viet Cong relentlessly attacked—but only at night, every night.

Spooky 71 was the gunship on duty that night. The troops on the ground called her Puff the Magic Dragon. The ship's three 7.26 mm miniguns could fire six thousand rounds a minute. More than twenty-one thousand rounds were on board. A burst from a Minigun looked like dragon's breath and sent a rain of fire toward the enemy. *Spooky*'s guns had already silenced two enemy mortars as the pilot, Major Kenneth Carpenter, turned the plane around to quiet a third.

Airman First Class John Levitow was on his 180th mission. It should have been his night off. He wasn't supposed to be on the flight, but Ken needed a loadmaster. So on February 24, 1969, it was A1C Levitow's job to grab a flare, set the timer, and hand it to the gunner. Then, standing in the open cargo door, the gunner attached it to a ten-foot lanyard, held the safety pin in one hand, and tossed the flare out of the plane.

The Mark 24 magnesium flares on board weighed twenty-seven pounds and were almost three feet long. Sixty were on board. A parachute kept the burning flare aloft for three minutes, lighting up the night sky. See the enemy, kill the enemy. *Spooky 71* was looking for its third kill of the night. But, at only one thousand feet up, *Spooky* was an easy target.

Just as the aircraft banked to make its turn, a North Vietnamese 82 mm mortar ripped a three-foot hole in the wing and peppered the fuselage with over thirty-five hundred pieces of shrapnel. One of the gunners, Sergeant Fuzie, was wounded in his back and neck but remembered seeing his three fellow airmen, Baer, Owen, and Levitow, instantly go down. Sergeant Baer was bleeding heavily. Later, Levitow would say he felt like he'd been hit by a 2 × 4.

It was Airman Owen who first realized they were in mortal danger: "I had the lanyard on one flare hooked up, and my finger was through the safety pin. When we were hit, all three of us were thrown to the deck. The flare, my finger still through the safety pin, was knocked out of my hand. The pin was pulled, and the flare rolled on the aircraft floor, fully armed."[1]

Pilot Major Carpenter fought to keep control. The plane finally went into a thirty-degree bank. From the intercom, he learned everyone in the back was injured. He also realized a live flare was rolling on the deck.

Airman Levitow had over forty pieces of metal in his legs and back. First, he came to the aid of a crewmember who was dangerously close to the open cargo door. Pulling him out of the doorway, he quickly looked for the flare. Nobody knew when the clock started. But, in ten seconds, a small explosion would deploy the parachute. In another ten, the magnesium would begin to burn at four thousand degrees. Next, the magnesium would burn through the metal floor and ignite the plane's fuel tanks. Finally, the explosion would kill the entire crew.

Nobody can explain why Airman First Class Levitow did what he did next. Maybe the loss of blood clouded his thoughts. He reached for the flare and missed. He grabbed again, missed again. With one last chance, he jumped forward, landing on the flare. He brought it to his chest. He could only crawl at this point, so he did. He crawled toward the open door with his arms around the live flare. Making it to the cargo door, he pushed it out, where it hit the prop wash and pulled away from the airplane. In an instant, it exploded, setting the night sky ablaze.

With two critically wounded crewmembers and over 3,500 shrapnel holes in the wing and fuselage, it's a wonder Major Carpenter made it back to Tan Son Nhut Air Base. But, looking in the back of his aircraft and seeing the trail of blood, it wasn't hard to understand what had happened. He would later say, "I'll never know. . . I'll never know . . . how Levitow managed to reach the flare and throw it out. In my experience, I have never seen such a courageous act performed under such adverse conditions."

Years later, Airman First Class Levitow was asked why he did it. The only thing he could think of was "temporary insanity."[2]

It took months for Airman Levitow to recover from his wounds. He went on to fly twenty more missions as a loadmaster in Vietnam—lighting up the sky in support of Army troops on the ground.

In August 1969, Sergeant Levitow was honorably discharged from the Air Force. His military awards included the Purple Heart and the Air Medal, with one silver and two bronze oak leaf clusters.

On May 14, 1970, President Richard Nixon presented the Medal of Honor to twelve US service members. Included were five members of the Army, three from the Navy, two Marines, one Air Force officer, and Airman First Class John L. Levitow, the lowest-ranking person to be awarded the Medal of Honor in the history of the Air Force.

John Levitow returned to the small town in Connecticut near where he was born. He worked for that state's Department of Veterans Affairs Administration for the next twenty-two years.

He would have preferred a quieter life. But fame didn't leave John alone. Instead, he became an icon. Veterans' groups wanted him to speak, as did civic groups around the nation. The Air Force wanted him back as well, not to reenlist but to motivate and teach.

Today, the John L. Levitow Award is presented to the top graduate in each of the airman, noncommissioned officer (NCO), and senior noncommissioned officer academies. You must graduate in the top 1 percent of your class to be eligible for the award. As important, if not more, you must be recommended by your classmates. It is the top award for each of the enlisted leadership schools.

I remember learning Sergeant Levitow's story when I made NCO in 1977. Today, every airman knows the story and legacy of Sergeant John L. Levitow.[3]

During the last years of his life, John fought cancer with the same determination that won him our nation's highest award.

John Levitow was inducted into the Airlift/Tanker Association Hall of Fame in 1998. A C-17 Globemaster III, *The Spirit of John L. Levitow*, was named in his honor. The aircraft now flies for the 105th Airlift Wing of the New York National Guard.

On November 8, 2000, John Levitow passed away. He is survived by his son, John Jr., of Charlotte, North Carolina; his daughter, Corrie Wilson, of Cromwell; Connecticut; his mother, Marion Levitow, of South Windsor, Connecticut; his sister, Mary-Lee Constantino, of East Hartford, Connecticut; and his grandson.[4]

Medal of Honor

CITATION:

For conspicuous gallantry and intrepidity in action at the risk of his life above and beyond the call of duty. Sgt. Levitow (then A1C), US Air Force, distinguished himself by exceptional heroism while assigned as a loadmaster aboard an AC-47 aircraft flying a night mission in support of Long Binh Army post. Sgt. Levitow's aircraft was struck by a hostile mortar round. The resulting explosion ripped a hole 2 feet in diameter through the wing, and fragments made over 3,500 holes in the fuselage. All occupants of the cargo compartment were wounded and helplessly slammed against the floor and fuselage. The explosion tore an activated flare from the grasp of a crewmember who had been launching flares to provide illumination for Army ground troops engaged in combat. Sgt. Levitow, though stunned by the concussion of the blast and suffering from over 40 fragment wounds in the back and legs, staggered to his feet and turned to assist the man nearest to him who had been knocked down and was bleeding heavily. As he was moving his wounded comrade forward and away from the opened cargo compartment door, he saw the smoking flare ahead of him in the aisle. Realizing the danger involved and completely disregarding his own wounds, Sgt. Levitow started toward the burning flare. The aircraft was partially out of control and the flare was rolling wildly from side to side. Sgt. Levitow struggled forward despite the loss of blood from his many wounds and the partial loss of feeling in his right leg. Unable to grasp the rolling flare with his hands, he threw himself bodily upon the burning flare. Hugging the deadly device to his body, he dragged himself back to the rear of the aircraft and hurled the flare through the open cargo door. At that instant the flare separated and ignited in the air, but clear of the aircraft. Sgt. Levitow, by his selfless and heroic actions, saved the aircraft and its entire crew from certain death and destruction. Sgt. Levitow's gallantry, his profound concern for his fellow-men, at the risk of his life above and beyond the call of duty are in keeping with the highest traditions of the US Air Force and reflect great credit upon himself and the Armed Forces of his country.

Sergeant John L. Levitow's grave is in Section 66 of Arlington National Cemetery. From the visitor center, take an immediate left on Eisenhower Drive. It's a little more than three-fourths of a mile. His grave is on the left.

The walk takes about seventeen minutes. When you get there, try to remember those twenty seconds.

John is with good company

CHAPTER 9

CALL ME NATE

Gravesite of Major General Nathan J. Lindsay

I WAS UNACCUSTOMED TO CALLING GENERAL OFFICERS BY THEIR first names even if they told me to, especially if I worked for them.

Major General Nathan J. Lindsay did tell me to, but I tried to ensure I wouldn't have to. Besides, isn't every senior officer's first name "Sir," even the female ones?

I didn't know Major General Lindsay much at all. I met him only twice. He was approaching his career's end while mine was just taking off. He had three master's degrees in engineering. I have a bachelor of arts in psychology. He was part of the Air Force's secret astronaut program. I had trouble landing a plane.

I wanted to understand a little about General Lindsay, so I asked an old friend who'd worked for him directly. My friend was one of a dozen full colonels who were General Lindsay's first-line managers of the organization. The other colonels managed the individual satellite acquisition programs with relatively small staffs of officers. My friend's officers flew the satellites and processed the intelligence. General Lindsay's other colonels had most of the money, while my friend had most of the people.

"What does a general officer do?" I asked.

"That's easy. Only two things. First, decide what's next. That means which program or capability we will need but can't yet imagine. More importantly, determine who's next. Decide which officer deserves a promotion or increased responsibility."

Because of colonels like my friend and others, only the best officers moved up. I'd say that only the brightest officers were advanced, but somehow, I got into that group as well.

General Lindsay controlled billions of dollars as the commander responsible for all the Air Force's secret spy satellite

22 Oct 89

Gary —

Your record of performance has been superb for your entire career. At your current location you have made a tough job look easy.

Keep it up — and let me help you find the right next job. Consider LA! I'd like to get you into a SPO.

Thanks for a great year.

Nate

A personal note

programs. His officers were an elite group. Each was specially selected and had one and often more engineering or science degrees.

Shirley Lindsay grew used to her husband being gone. His trips were often unexpected and never announced, so she knew little of what he did. What she did know was simple: if Nate was traveling, so were many officers who worked for him. Shirley knew the wives and families and tended to them as Nate tended to his officers. Because of Nate and Shirley, the organization was more a family than the military.

Nate began his career in 1959 when it seemed like every service member was being shipped to Vietnam. Nate stayed stateside to manage ultrasecret satellite programs that provided key information from what we called "the high ground." Anyone can go to war, but a few visionaries must stay behind and focus on the new systems designed to do only two things: find the bad guys and save the lives of our troops. General Lindsay knew his work was important, even if Shirley and their children could not.[1]

War is not an exact science. Things go wrong. If a surgeon makes a mistake, a patient can die. During a war, if a commander makes a mistake, many people can die. In every profession, better information leads to better decisions. General Lindsay lived his career ensuring commanders had the best information when it could do them the most good.

Major General Nathan J. Lindsay retired and passed away before the formation of today's United States Space Force. He would be proud to know that many officers he mentored, groomed, and promoted now lead our nation's newest military branch. Nate grew today's Space Force leaders. Military personnel in US Space Force are called *guardians*—a fitting title.

Today, General Lindsay rests with the one thousand Soldiers, Marines, Sailors, and Airmen who lost their lives serving in Afghanistan and Iraq. Nate is a guardian of those heroes. I know he will continue to care for each of them as he did for me.

Major General Lindsay's gravesite is at the northeast end of Section 60—almost three-fourths of a mile from the visitor center. Take a left on Eisenhower Drive and then turn left on Sergeant Alvin C. York Drive.

Major General Nathan Lindsay, official USAF photo

DEVOTION

There is nothing in the world like the devotion of a married woman. It is a thing no married man knows anything about.

—Oscar Wilde, *Lady Windermere's Fan*

CHAPTER 10

MAMA JOE'S TABLECLOTH

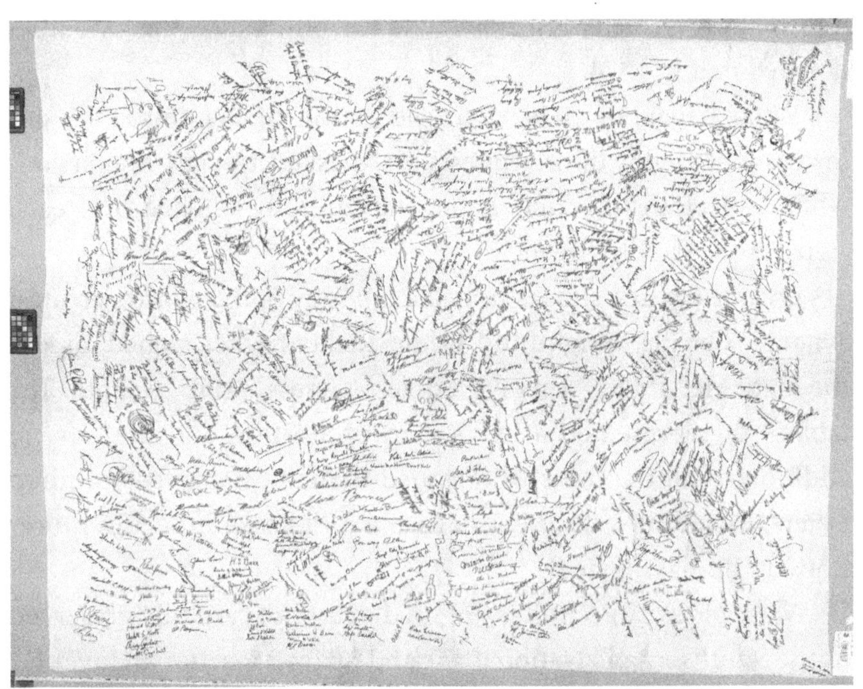

Mama Joe's Tablecloth, Courtesy of the Smithsonian Air and Space Museum

Mama Joe entertained, cooked, and housed Jimmy's guests for six decades. He was an aviation pioneer and war hero. She was the friendly soul, the welcoming warmth after a long day.

In a previous chapter, I detailed the story of General James H. Doolittle and his daring raid on Tokyo, Japan, during World War II. I also introduced his beloved wife of seventy-one years, Josephine Daniels Doolittle. I wrote subsequently to the Smithsonian's National Air and Space Museum asking for information on her most notable tablecloth, which the Doolittles donated to it in 1974.

What began in 1929 as a celebration of her husband's all-instrument flight, dinner guests were asked to sign Mama Joe's white tablecloth in pencil. Later, she carefully stitched over each signature with silk thread.

Over the years, she fashioned a tapestry of the world's greatest aviators. Orville Wright and Eddie Rickenbacker signed, as did the first chief of the Army Air Forces, General "Hap" Arnold. The tablecloth represented a who's who of aviation and the arts. In addition to all the fliers who signed the tablecloth, five hundred others, including author Laura Ingalls Wilder and the adventurer Lowell Thomas, signed.

Mama Joe's tablecloth is not on display but carefully preserved in the Smithsonian's National Air and Space Museum archives for future USAF airmen, US Space Force guardians, and all who love the vastness of the sky.[1]

For completeness, here are verses III and IV of the US Air Force song. It's okay if you don't remember. I'll help you.

(Verse III)
Here's a toast to the host
Of those who love the vastness of the sky,
To a friend we send a message of the brave who serve on
high.
We drink to those who gave their all of old
Then down we roar to score the rainbow's pot of gold.
A toast to the host of those we boast, the US Air Force!
(Verse IV)
Off we go into the wild sky yonder,
Keep the wings level and true;
If you'd live to be a grey-haired wonder
Keep the nose out of the blue!
Fly to fight, guarding the nation's border,
We'll be there, followed by more!
In echelon we carry on.
Oh, nothing'll stop the US Air Force!

Josephine Daniels Doolittle is buried with her husband at Arlington National Cemetery in Section 7A, just below the Tomb of the Unknowns—a little over half a mile from the visitor center.

Josephine Daniels Doolittle is buried with her husband

CHAPTER 11

PAMELA OPAL LEE

Gravesite of Pamela Opal Lee Murphy

One Monday after Easter Sunday, we finally got the chance to visit Pamela Opal Lee's grave. A Monday drive in the rush hour traffic of Los Angeles is not for the faint of heart. But seeing her grave was on my mind. I had to complete the story I'd started several years before. So off we went. My bride, Lillian, and I were taking yet another drive to find one more tiny sliver of America.

Forest Lawn Memorial Park in Hollywood Hills is enormous. We drove through the gates and blindly followed the GPS. We must have missed the information center. After a few miles, the road ended. We stopped at the parking lot of a large building. The Court of Liberty housed statues of George Washington and Thomas Jefferson. It also has the largest glass mosaic in the United States, depicting twenty-five scenes of early America. Unfortunately, the hall was under renovation, and no one was around to give us directions.

Forest Lawn in Hollywood Hills opened in 1952 and houses almost 120,000 graves. Altogether, Southern California has eleven Forest Lawn cemeteries. If you're looking for a famous actor or actress, they're probably buried at one of the Forest Lawn cemeteries. The graves of Humphrey Bogart, Mary Pickford, Walt Disney, Clark Gable, and other celebrities dot the rolling green lawns. The grave we searched for was none of the above. Neither fame nor fortune had befallen Pamela Opal Lee.

The rolling hills overlooking the movie studios were still full of families and small children carrying flowers and foil balloons. Easter Sunday was like one big picnic at the cemetery. A sea of flowers covered most of the graves. The graves at Forest Lawn were row upon row of metal markers in the ground with no traditional headstones. From a distance, the grass-covered hills look more like a city park than a cemetery. It was Monday, and the dozens of families dotting the hillsides were just the remains of the crowd left over from Easter Sunday.

Walking out of the Court of Liberty, I had no idea where to find Lee or even whom to ask. Realizing we were lost, a maintenance worker named Nathan finally took pity on us and showed us the section markers. The small, round concrete medallions were not much bigger than a US geological survey marker. You had to move the grass to see them. Nathan motioned for us to get back in the car and follow him. I guess whatever chores he had could wait. Nathan and the dead had learned to be patient.

We followed Nathan for two miles into the park's practically deserted older area. Only one or two families shared our hillside, littered with unattended graves. Finally, Nathan stopped his maintenance truck, and we walked down a hill covered with graves to find Section 2793, Space 1. The section markers were hard to find and even harder to read. We would never have found her on our own.

Pamela Opal Lee's twenty-year marriage to Audie Murphy was underscored with problems. Their romance began in a whirlwind. He had just finalized his divorce from a Hollywood glamour queen. In 1951, Pamela Opal Lee was an airline stewardess. He was a budding actor. They would go on to have two sons. Terry Michael was born in 1952, and James Shannon was born in 1954. Her husband's acting career took off, but his gambling grew worse. With his income from acting, he raised and raced quarter horses. Together, they owned a comfortable house in Van Nuys and ranches in Arizona and Riverside, California.

Murphy became addicted to sleeping pills. His nightmares were so real that he slept with a loaded revolver under his pillow. He sometimes wouldn't come home for days and was arrested in Burbank for assault and battery. They say he fired a gun at someone, but those charges were somehow dropped. Murphy spoke candidly about the stress of combat.[1] He championed improving the care that veterans should receive for post-traumatic stress

(PTS). He recognized his problems and tried to heal the best he could. He quit the sleeping pills cold turkey, and he never developed a dependency on alcohol. He'd refuse when tobacco or whiskey advertisers approached him to market their products. He knew his status as a hero meant he was a role model—he tried to live up to it.

Bad investments and a mountain of debt had the IRS investigating him. With all of his problems, Pamela never abandoned him. He always remained her hero. Then, in 1971, before the government could make a case against him, he died in a small plane crash in the foothills of Virginia's Appalachian Mountains. She would live for thirty-six more years. She never remarried.

After he died, his gambling debts ruined Pamela, forcing her and the children to sell their home and move to a small apartment. She took a job greeting patients at the Sepulveda VA Medical Center. She went on to have a thirty-year career working as a patient advocate. Pamela worked for years to pay off his debts.

At the VA, she'd say every veteran was a hero and should be treated with respect. She worked serving veterans until she retired in 2007. Known as the lady with the clipboard, she was often reprimanded for cutting the VA's red tape. If she saw veterans waiting too long, she'd march them directly into the doctor's office. Those were her soldiers.

"Nobody could cut through VA red tape faster than Mrs. Murphy," said veteran Stephen Sherman, speaking for thousands of veterans she befriended over the years. "Many times I watched her march a veteran who had been waiting more than an hour right into the doctor's office. She was even reprimanded a few times, but it didn't matter to Mrs. Murphy. Only her boys mattered. She was our angel."[2]

At one point in her career, some of the VA's higher-up bureaucrats determined that her job was excess and needed to be eliminated. A protest ensued, with hundreds of workers and friends demonstrating at the entrance to the hospital. After that, the VA managers changed their minds.

Initially, nobody knew who her husband had been—she was just the friendly face who greeted them and guided them through the hospital maze. Then, when word got out about her husband, wounded vets cried and rushed to hug her. Many just wanted to touch her.

"He was my hero," they'd say.

"You're the hero," she'd reply.

Pamela Opal Lee grew to become a legend at the VA. Her first few years were about him. Her last thirty were all about the veterans she served. She was their angel.

On April 8, 2010, Pamela Opal Lee Archer Murphy died at her home in Canoga Park, California. She was eighty-six years old. Her husband, Major Audie Murphy, died thirty-six years earlier. He was the most decorated soldier of World War II. He was awarded the Medal of Honor and every other award for valor the United States can bestow. He was also awarded five more medals for bravery from France and one from Belgium.

Audie Murphy's grave is one of the most visited sites at Arlington National Cemetery. He was buried with full military honors. The Ambassador to the United Nations, George H. Bush, Army Chief of Staff, General William Westmoreland, and hundreds of soldiers attended his funeral.[3]

Thirty-nine years later, Pamela Opal Lee Archer Murphy's funeral packed the Forest Lawn Memorial Park chapel. Hundreds of the veterans she had comforted over her thirty-five years at the

VA came to say goodbye to their angel. There were no flowers on her grave when we visited. Easter Sunday had come and gone.[4]

On Memorial Day, soldiers from the 3rd Infantry Regiment (The Old Guard) will place a small American flag at Audie Murphy's grave. The flag will be precisely one boot length away from his headstone. The soldiers will continue placing flags at the graves of all the US service members buried at Arlington.

On the Sunday after Memorial Day, Lillian and I returned to the grave of Pamela Opal Lee Archer Murphy. I placed a small American flag, and Lillian brought some flowers.

Thank you for your service, Pamela Opal Lee.

Gravesite of Major Audie L. Murphy

CHAPTER 12

A BLANK CHECK

VA outpatient clinic, Sepulveda, California

As I mentioned, Lillian and I got the chance to return to the grave of Pamela Opal Lee Murphy. We had vowed to place a few flowers and a flag at Pamela's grave for Memorial Day.

The roses were from our front yard. The flag was on sale at Lowe's.

Nobody works for the Department of Veterans Affairs for the money. Pamela worked there for thirty-five years. Her face was often the first that veterans saw when they entered the hospital. Sometimes, it was the last one they'd ever see. She saw the pain of PTS every day she went to work. She had felt that pain years earlier during her marriage to Audie Murphy.

After World War II, Audie Murphy spoke freely about the stress of combat and advocated for better mental health care for Soldiers returning from Korea and Vietnam. Pamela Opal Lee raised two sons and tirelessly worked helping disabled vets at the Sepulveda VA Medical Center.[1]

It was a beautiful day in Southern California when we visited her grave. Too beautiful a day to spend at home. We wanted to make a day of it, so we drove to see Audie Murphy's star on the famous Hollywood Walk of Fame. Hollywood and Vine are all of five miles as the crow flies, but about two hours if you drive. No matter where you start, driving into Hollywood is both a career and an adventure.

Audie Murphy's film career spanned twenty-one years, including forty feature films and a television series. In 1945, after reading about Audie's heroic acts in *Life* magazine, the actor James Cagney brought him to Hollywood, where Audie played the leading role in Westerns and starred as the youth in the screen adaptation of Stephen Crane's classic, *The Red Badge of Courage*. In 1949, Audie Murphy authored his best-selling autobiography, *To Hell and Back*. Then, in 1955, he portrayed himself in the movie version.

Walk of Fame star at 1601 Vine Street, Hollywood, California

He suggested Tony Curtis play the leading role, but the studio insisted on him. The public loved it, but the reviews were mixed. At the time, the movie was the second-highest-grossing box office hit for Universal Studios.

Audie Murphy was also an accomplished songwriter and poet. He loved country and western music and wrote "Shutters and Boards" and "When the Wind Blows in Chicago." Roy Clark, Dean Martin, Teresa Brewer, Bobby Bare, and Eddy Arnold lent their voices to his music.

His autobiography, *To Hell and Back*, contained this poem:

"The Crosses Grow on Anzio"

Oh, gather 'round me, comrades; and
listen while I speak
Of a war, a war, a war where hell is
six feet deep.
Along the shore, the cannons roar. Oh
how can a soldier sleep?
The going's slow on Anzio. And hell is
six feet deep.
Praise be to God for this captured sod that
rich with blood does seep.
With yours and mine, like butchered
swine's; and hell is six feet deep.
That death awaits there's no debate;
no triumph will we reap.
The crosses grow on Anzio, where hell is
six feet deep.
—Audie Murphy, 1948

On Friday, June 10, 2022, Lillian and I visited the VA hospital in California's San Fernando Valley, where Pamela Opal Lee had worked for many years. The original hospital was constructed in the 1950s to care for thousands of WWII and Korean War veterans settling in the area. Southern California was booming, and the GIs needed a hospital, so in 1952, Lester and Mary Gentry donated 160 acres of their land for the hospital. In time, the hospital treated over 275,000 veterans every year.[2]

Then, in 1994, the Northridge earthquake hit, and most hospital buildings were damaged. It would take hundreds of millions of dollars to repair, so the VA decided to close the inpatient hospital and build a new outpatient clinic. Opposition to closing the hospital was intense. Local congressional representatives eventually sided with the VA. Veterans who had used the Sepulveda Hospital for years were told they would have to endure a two-hour commute in bumper-to-bumper traffic to the VA hospital in West Los Angeles. The West Los Angeles complex is huge: it's the fourth largest VA hospital in the United States, and its seven hundred beds needed to be filled by the Sepulveda VA patients.

Argo, the 2012 Best Picture of the Year, was filmed on the Sepulveda VA hospital grounds. The hospital and the US embassy in Tehran apparently look remarkably similar. In addition, the new outpatient clinic was used as Seattle Grace Hospital for the *Grey's Anatomy* television series.[3]

Earthquakes and movie sets are a byproduct of living in Southern California. However, we planned to visit the hospital for another reason. We wanted to find anyone still around who might have known Pamela Opal Lee. But Pamela had retired in 2007, over fifteen years ago.

Lillian and I visited the hospital and asked everyone we could talk to—the security guard, the receptionist in the main lobby, random nurses, and techs. One person led to another, and we were finally introduced to Kelli Dela Milera, who was Pamela's friend and worked in the Employee Health Department. We could have asked a few questions and called it a day, but I wanted to get the VA's official permission for my interview. The senior person of the Sepulveda VA was Charles Green, the site manager, but he couldn't approve my request. To get permission to interview Kelli or anyone else in the VA, I'd have to talk to the VA's public relations department in downtown LA—so I did. Now many of you are thinking just what Lillian often thinks: "There goes Gary, tilting at windmills again."

Maybe I was. But people are just people. The VA is a vast bureaucracy and predisposed to the answer *no*. But I wasn't going no-fishing, and I wanted to meet Kelli to learn what she knew about Pamela Opal Lee.

The VA bureaucracy took less than a week. Charles immediately forwarded my request, and someone in the public affairs office got back to me a few days later. Bureaucracy or not, people are just people.

The day finally came to meet Ms. Kelli Dela Milera officially. In 2006, during Kelli's first year working at the VA, she met Pamela. Pamela retired a year later. Every morning they'd exchange greetings. They were assigned to different clinics, but their paths often crossed. Pamela's reputation was well known. She wanted the best for the vets who came through and wasn't shy about helping them get it. Pam's attitude seems to have rubbed off. Kelli would say, "Damn, the bureaucracy. We've got veterans here who need our help." I had a few minutes to sit down with Kelli and learn what she knew about Pamela Opal Lee.

"Kelli, nice to see you again. Thanks for taking the time to speak to me."

"I knew you'd be back. So, I'm happy to help in any way I can."

"Why do you do this? I mean, why do you work at the VA?"

"You all [veterans] wrote me a blank check and gave it to me. A blank check with your service. A check I try and repay every day."

"Can I give you a hug?"

Kelli lives with her four twenty-pound cats, Poopahlee Pooh, Stevie Norman Janowski, Fiona Norma (dictator of the world), and Nelson Mandela. She also has a dog, Carrie—a greyhound-chow mix. In eight years, Kelli will retire from the VA. She will repay that blank check for eight more years. Then, she plans on buying an RV and traveling around the country.

Thank you for your service, Kelli Dela Milera. A hero serving heroes every day.

Ms. Kelli Dela Milera

CHAPTER 13

LENA

Sergeant Lena Mae Basilone USMC, widow of John Basilone, prepares to christen the destroyer USS Basilone, December 21, 1945 (Photo Courtesy US Navy)

Lena Mae Riggi Basilone was the daughter of Italian immigrants and grew up in Portland, Oregon. After high school, she went to business school, but when World War II started, Lena joined the Marines and served as a field cook.

The Marines needed cooks, lots of them. Lena was stationed at Camp Pendleton just north of San Diego. She was thirty years old when she met John Basilone. He had just returned from the Pacific. He was a hero.

Fresh off a savings bond drive, he'd helped raise over a million dollars for the war effort. He dined with the president and flirted with movie stars. The brass wanted him to accept a commission. He turned them down: "I ain't no officer and I ain't no museum piece. I belong back with my outfit."[1]

They would have been married fifty-five years had he lived until she passed away. She lived eighty-six long years. John, just twenty-eight short ones.

In the United States Marine Corps, Gunnery Sergeant John Basilone is a legend. City streets, parks, and two Navy destroyers are named after him. He was awarded the Medal of Honor for leading a machine-gun battery and holding off thousands of Japanese soldiers on Guadalcanal. When his ammunition ran out, he used his pistol and a machete to repel the remaining enemy.

He returned to the States to train new Marines when he met Lena. They were married on July 10, 1944, at Saint Mary Star of the Sea Catholic Church in Oceanside, California. John and Lena honeymooned near her childhood home in Portland, Oregon. But the war wasn't over, so the Marines sent him back to the Pacific.

On the first day of the Battle of Iwo Jima, as a machine-gun section leader, he led a company of Marines, advancing on a heavily fortified blockhouse. His fellow Marines were pinned down

Saint Mary Star of the Sea, Oceanside, California

and taking fire. He went along the side and onto the top of the enemy blockhouse, destroying the fortification.

Next, Sergeant Basilone aided one of our tanks pinned down and trapped in a minefield. They were taking heavy fire from mortars and enemy artillery. He guided the tank out of the field and to a position of safety. Gunnery Sergeant John Basilone was killed that same day. For his actions, he was awarded the Navy Cross, the Navy's second-highest award, bestowed for valor.[2]

John Basilone died in 1945—just seven months after their wedding day. Lena learned of John's death on February 19, 1945, her thirty-second birthday. He was the love of her life. She never remarried. According to her friend Barbara Garner, Lena would say, "Once you had the best, you can't settle for less"[3]

Gravesite of Gunnery Sergeant John Basilone

Lena lived out her remaining years in Southern California. She bought a small house in Lakewood and loved to cook for friends on special occasions. She shunned publicity and rarely spoke in public.

She worked for the local electric company and volunteered for the Woman's Marine Association and other veterans' groups.

On June 11, 1999, at age eighty-six, Lena died peacefully at her home. Three days later, a seventeen-mile stretch of Interstate 5 was dedicated to her husband, Gunnery Sergeant John Basilone. Unfortunately, she never got to see the ceremony. Unlike her husband, no roads are dedicated to Lena Basilone.

Lena didn't want to be buried at Arlington next to John. The Department of Veterans Affairs and the Marine Corps wanted to bury her next to John, but she refused, saying, "I don't want to cause trouble for everyone."[4] She still wore the wedding ring John gave her fifty-five years earlier.[5]

Gunnery Sergeant John Basilone was buried at Arlington National Cemetery with full military honors. To visit his grave, take the second left out of the visitor center. His grave is a little over half a mile walk in Section 12.

Lena Mae Riggi Basilone is buried in Riverside National Cemetery in Riverside, California. Her grave is in Section 50, Site 5557.

Gravesite of Lena Basilone

CHAPTER 14

MEDGAR AND MYRLIE

Gravesite of Technician Fifth Grade Medgar W. Evers

I showed up at Lackland AFB in San Antonio, Texas, in early August 1974. The Vietnam War would last only six more months. After nineteen years of fighting, our troops were coming home, and I was just beginning my military career. I had successfully survived the lottery. The draft ended before my US Selective Service number was called, yet I volunteered to serve my country.

Then, in April 1975, I watched in disbelief as Saigon fell. I felt embarrassed as our troops and embassy personnel clambered to find a way out. I was newly married and at my first duty station in Blytheville, Arkansas. Too new to understand our adopted military community, Lillian and I were also too "foreign" to join or be accepted by the residents of a small rural Southern town. All I knew was that we were different, out of place, and alone. Southern California and our high school friends seemed a million miles away.

The first conversation I had with a Black person was in basic training. Sure, I'd competed in high school sports with other schools that had Black athletes. But I had never talked with an African American. We might as well have been from different countries. At nineteen, I became fascinated by how people's perspectives influenced their views of the same facts. I observed a young sergeant, a white guy who grew up dirt poor in Mississippi, speak with joy when his third-grade class was dismissed early because President Kennedy was assassinated. His words held me in shock, and I froze. He remembered November 22, 1963, as a time of celebration. I remember that day with profound sadness.

As hard as I try to understand the perspective of someone whose life experiences are notably different from mine, I find myself failing to truly comprehend. As much as I believe I can see injustice, prejudice, and inequalities, I am, after all, white, male, and, at times, tone-deaf.

As I approached Section 36, looking for the gravesite of Sergeant Medgar Evers, the inequalities that still exist in our society and how various perspectives shade our facts flooded my mind. Remembering those deep philosophical discussions I had with the first friends I made in the Air Force, those few Black Airmen, a long way from home just like me, made me realize that we can only see with our own eyes and tell the truth from our experiences.

Sergeant Joel Johnson was soft-spoken, or at least he talked in almost a whisper. His Southern Mississippi accent often defied my understanding. If he smiled at the end of a sentence, that was my cue to laugh. Sergeant Thomas Perkins was from Detroit and tried hard to be one of the good old Southern boys. Even I knew he was a faker and didn't fit in. Don't get me wrong, everyone liked Tom. He was a good guy. But you are either born in the South or not, Black or white. Staff Sergeant Alfred Harrison was our boss, our noncommissioned officer in charge. Al was from Jackson, Mississippi, and was going through a terrible divorce. His wife was getting everything, including their nine-year-old son. Al should have lived off-base. Instead, he lived in the airmen's barracks. He needed the money.

As the years roll by, I believe Joel, Tom, and Al would've liked this story. They wouldn't be able to relate to me and my musing, and they wouldn't be able to connect to Myrlie and Medgar Evers either. Joel, Tom, and Al weren't heroes of the civil rights era. They lived in the immediate aftermath. They were victims. They would also like this story because they didn't want or need another Black friend. What they accepted in me was something they had little of, somebody white to talk to. This story would only cost them a little time. We seemed to have a lot of time back then.

I was at Lowe's the other day. I like Lowe's. The workers are friendly, and they know what aisle everything is on. Lillian wanted some floating shelves for the bathroom to display all the stuff that should be boxed up and put away. I like Lowe's because they give me a discount. I like discounts. Being retired from the military, I get a 10 percent discount at Lowe's for buying stuff I essentially don't need.

IHOP gives active duty and retired service members a discount too. They will give you your fifth pancake for free. I wish more merchants gave discounts to veterans. Did you know that if you're a 100 percent disabled veteran, you get a special license plate that allows you to park up front at most businesses? In some states, you also can get a free fishing license. Best of all, in some states, you pay no property tax. Now, that's cool.

Of course, if you have a 100 percent disability rating from the VA, whatever injury you sustained on active duty will never get better. It usually means you can't work and often cannot care for yourself.

I like that injured veterans who served our nation get some special benefits. Injured veterans should get benefits beyond what your typical service member gets. I don't think we can ever thank disabled veterans enough. But how do we repay a veteran who lost a limb or worse? I guess we can't. We can never compensate them for their sacrifice.

I'll take my 10 percent off at Lowe's. I'll pay for my own fishing license, but then again, in my twenty-two years on active duty, I was never injured.

Joel would like this next part. Being from Mississippi, Al would lie and say he knew Myrlie. Nobody would listen to what Tom had to say.

Myrlie Louise Beasley was born on March 17, 1933. As of 2023, she's ninety years old. She's outlived two husbands. She's devoted her entire life to the cause of civil rights. Her first husband, Medgar, was a field secretary for the National Association for the Advancement of Colored People (NAACP), and she worked by his side.

In 1950, on the first day of college at Alcorn State University, she met and fell in love with Medgar Evers, an Army veteran eight years older than her. They married on Christmas Eve 1951 and would have three children. Medgar graduated from Alcorn State in 1952. In 1954, they organized voter registration drives and worked for integration after the US Supreme Court struck down segregation in public institutions in its decision in *Brown v. Board of Education*. With integration guaranteed by the highest court in the land, Medgar applied to the University of Mississippi Law School. He would be the NAACP's test case. Because of his race, Medgar Evers's application was rejected. Of course it was. This was 1954 and Mississippi.

Undeterred, Medgar and Myrlie continued to organize boycotts of whites-only businesses and public places.[1]

The Everses were frequent targets of death threats. In 1962, after organizing a boycott against white merchants who didn't allow Black customers, their home in Jackson, Mississippi, was firebombed. On Wednesday, June 12, 1963, Medgar Evers was shot as he got out of his car at home. He bled to death in his driveway in front of his wife, Myrlie, and their children. He still held his car keys in one hand, and the other clutched a few T-shirts with the slogan "Jim Crow must go." Medgar Evers was thirty-seven when he died.

Here are a few memorable quotes by Medgar Evers:

"You can kill a man, but you can't kill an idea."

"When you hate, the only person that suffers is you because most of the people you hate don't know it, and the rest don't care."

"The Negro has been here in America since 1619, a total of 344 years. He is not going anywhere else; this country is his home. He wants to do his part to help make his city, state, and nation a better place for everyone, regardless of color and race."[2]

Myrlie Louise Evers-Williams continued Medgar's work for social justice. She married her second husband in 1976, Walter Williams, a civil rights activist who studied Medgar and Myrlie's work.

Myrlie was the first Black woman to serve as Commissioner of Public Works for the City of Los Angeles. In 1995, the year her second husband died, she was appointed chairperson for the board of directors of the NAACP. She gave the invocation at President Obama's inauguration in 2013—the first African American woman ever to do so. Medgar and Myrlie's home in Jackson, Mississippi, became a national monument.

Sergeant Medgar Evers served in the US Army from 1943 until he was honorably discharged in 1945. He served in a segregated unit supporting combat troops after D-day.[3] He was buried with full military honors at Arlington National Cemetery. His grave is in Section 36, a short five-minute walk from the visitor center.

A known white supremacist was arrested for Medgar Evers's death in June 1962. The killer was released from jail after an all-white jury failed to reach a verdict. Myrlie Evers-Williams, however, never gave up seeking justice for her husband's assassination. She waited for three decades until a new county judge was appointed. KKK member Byron De La Beckwith was finally convicted of Medgar's murder based on new evidence.[4]

If you served in the military supporting our troops reclaiming Europe and devoted your life to social justice, you should get some sort of discount at stores and such.

I don't think you should be shot in the back.

Medgar and Myrlie Evers' Home in Jackson, Mississippi
(cc-by-sa-3.0-migrated-with-disclaimers)

MONUMENTS

What you leave behind is not what is engraved in stone monuments, but what is woven into the lives of others.

—Pericles

CHAPTER 15

IT'S NOT JUST A JOB

USS *Maine* Memorial

IN 1976, BATES ADVERTISING DEVELOPED ONE OF THE LONGEST-running slogans ever: "Navy. It's not just a job. It's an adventure." Over the years, I've had a few bad jobs. The worst ones mostly were when I was young, before the military, and didn't know any better. I remember in my high school years spending a night unloading fruit from shipping containers down at the pier in San Pedro. Luckily, I did this for just one night. My friends and I blew much of our earnings on a "king's breakfast" at Norm's Diner.

I also remember helping my friend Gary place railroad ties on the side of someone's hill. Gary worked for a guy who owned a few houses and always had an odd job or two to keep us busy. I can't imagine carrying a railroad tie up the side of a hill anymore. I remember spending the following day using kerosene to get the creosote off my hands.

For a short while in elementary school, I had a paper route. And sometimes I would fill in for a friend who also had a route. I couldn't always remember the right house, and many people complained. Getting up at four in the morning wasn't my thing back then. I didn't last long.

Who knows why anyone joins the military? A common myth about why youngsters enlist is to escape little to no job opportunities in their hometowns. Some are encouraged to go into the military to grow up. Another misperception for voluntarily joining the military is to avoid jail time. In high school, I was a C student with limited job prospects, with thoughts of going to a junior college at night. For me, joining the military seemed like the least bad option.

Today's military is nothing like the military of 1898. The military does not take at-risk youths or those with a criminal record. While young people might join to find themselves, as I did, there are no do-overs if you mess up. Long gone is the principle of three strikes and you're out. In today's military, it's one strike and you're gone. For example, smoking a joint or getting a DUI would be grounds for immediate discharge.

When you walk around Arlington, it is hard to miss the memorial to the USS *Maine*. The mast is taller than anything around it. When I read the names of some of the Sailors who died when the ship exploded, I couldn't help but think of the clever advertising that the military does to entice young men to enlist. Often the actual jobs they get are nowhere near the hype sold to them by a recruiter.

In 1898, the peasants in Cuba were starving. A civil war raged everywhere on the island. Cuba was once a stronghold of Spanish colonialism. The Cuban people wanted the Spaniards out, as did the United States, but President McKinley didn't want to become involved in a war in the Caribbean.

Major newspapers of the day fed the average American citizen a steady diet of yellow journalism and pushed the country toward war. The papers exaggerated the plight of the peasants and often didn't bother with facts to make their argument. McKinley reluctantly sent the USS *Maine* to Havana Harbor as a show of force and American firepower. This move was explained to the American people as necessary to protect American interests.

The USS *Maine* was a steel-hulled battleship commissioned in 1895. It carried a crew of 355 officers and men. Two triple-expansion steam engines fed by eight coal-burning boilers powered her hull of 324 feet. She had a top speed of sixteen and a half knots.[1]

On the night of February 15, 1898, a massive explosion ripped through her hull. The coal and boilers were at the bow, the site of the detonation. A total of 261 lives were lost. Of the 94 survivors, only 16 escaped uninjured. Most of the officers were quartered near the stern and survived.[2]

Everyone immediately blamed the Spanish. A torpedo? A mine? Journalists quickly coined the phrase "Remember the

Maine, to Hell with Spain." McKinley was then forced to declare war. Several investigations into the explosion began. To increase the speed of the Navy's steam-powered ships, naval engineers had switched the type of coal to one that burns hotter but outgasses methane. In 1974, Admiral Hyman Rickover finally did his own investigation and determined that a spontaneous explosion in the ship's coal bins sunk the USS *Maine*.[3]

Because the explosion occurred at the bow of the ship beneath the enlisted quarters, almost all of the 261 killed were enlisted men. Only two were officers.

"Join the Navy and see the world." But the entire world looks the same from a ship's boilers and engine room. "Navy. It's not just a job. It's an adventure." Tell that to Ordinary Seaman William Gorman or Coal Passers Frank Gardner, Patrick Grady, John Fougere, and James Furlong. These are just a few of the many men lost.

To visit the USS *Maine* memorial, at Arlington, continue straight from the visitor center on Roosevelt Drive. Next, turn right on Memorial Drive and then make a sharp right on Farragut Drive. Make a sharp left on Sigsbee Drive. You will see the mast of the USS *Maine* long before you arrive.

Names of some of the lost souls of the USS *Maine*

CHAPTER 16

ANOTHER DAY THAT WILL LIVE IN INFAMY

9/11 Pentagon Memorial marker

MOST OF OUR LIVES ARE BLANKETED UNDER BLAND, NONdescript days and empty, uneventful nights. What did you do last Tuesday? Probably the same thing you did every Tuesday a hundred times before. Perhaps you remember Sundays as special. Sundays are for church or maybe sharing a meal with family and friends. I like Sundays. They seem warmer than the rest of the week.

Sometimes, not often, maybe only once in a lifetime, a single day emerges and burns itself into our consciousness. Abraham Lincoln was shot on a Friday. Pearl Harbor was attacked on a Sunday.

President Franklin Delano Roosevelt told the world by radio: "Yesterday, December 7, 1941, a date which will live in infamy, the United States of America was suddenly and deliberately attacked by naval and air forces of the Empire of Japan."

On another Sunday, April 14, 1935, the worst sandstorm in American history devastated the plains of the Midwest. The Dust Bowl would last for four more years.

In 1963, CBS news anchor Walter Cronkite read these words on air: "From Dallas, Texas, a flash—apparently official—President Kennedy died at 1 p.m. Central Standard Time." November 22, 1963, was a Friday.

On Tuesday morning in 2001, most of America was just waking up. It was another beautiful September day when the first plane hit the North Tower of the World Trade Center. Then the South Tower was hit. A third plane hit the Pentagon. A fourth plane was headed for either the White House or the Capitol but crashed in a field in Pennsylvania. The hijackers' plans were thwarted by a few passengers. We were now at war.

President George W. Bush spoke to the American people:

> I want you all to know that America today, America today is on bended knee, in prayer for the people whose lives were lost here, for the workers who work here, for the families who mourn. This nation stands with the good people of New York City and New Jersey and Connecticut as we mourn the loss of thousands of our citizens. . . . I can hear you! I can hear you. The rest of the world hears you. . . . And the people who knocked these buildings down will hear all of us soon.

Beginning at 7:59 a.m., four commercial airliners from three different airports took off over the next forty-three minutes.

Beginning at 8:46 a.m., the four commercial airliners were deliberately crashed one by one. Two planes flew into the Twin Towers in New York City, one into the Pentagon, and one into a field near Shanksville, Pennsylvania.

If you were alive on that day, you know what you were doing on Tuesday morning, September 11, 2001.

In Washington, DC, flying low to avoid the series of high-rise buildings in Rosslyn, Virginia, was United Airlines Flight 77.

At 9:37 a.m., Flight 77 crashed into the west side of the Pentagon. All fifty-three passengers and the six crewmembers died instantly. As a portion of the building's outer ring, known as the E-Ring, collapsed, 125 Army, Navy, and civilian civil servants lost their lives.[1]

By 9:45 a.m., the vice president ordered, and then the president reordered, the halting of all air traffic across the country. Every plane in the air was directed to land at the nearest airport. International flights inbound for the United States were diverted to Canada or Mexico.

At 9:59 a.m., Father Mychal Judge, chaplain to the FDNY, stationed at an emergency command post in the lobby of the North

Tower, was struck by falling debris from the collapsing South Tower. He died of blunt-force trauma to the head. He's listed as victim number one.

By 10:00 a.m., the passengers of hijacked United Flight 93, the fourth plane, realized they had little chance of survival. Receiving cell phone calls from family and friends, the passengers learned the fate of the first three planes. Their aircraft was now a guided missile—a bomb. With a flight attendant, the pilot, and the copilot dead, a few passengers tried to overtake the hijackers. The terrorists at the controls violently swung the plane left and right. The passengers used a cart to rush the cockpit door. The aircraft pitched into a steep dive just before they broke open the door. America had begun to fight back.

At 10:03 a.m., Flight 93 crashed into the Pennsylvania countryside. All thirty-three passengers and seven crewmembers on board died instantly. The intended target was the US Capitol or the White House.

By 10:15 a.m., fighter planes scrambled from New Jersey and Long Island with orders to shoot down any suspicious aircraft that didn't respond to radio calls. The order included commercial airliners with US citizens onboard.

On that sunny Tuesday morning, on what should've been a beautiful day in September, thousands of firefighters from over seventy-five station houses throughout New York's five boroughs responded, and 343 first responders rushed into the smoke and the collapsing towers, never to return. They ran into the fire.

Before the towers collapsed, the people trapped inside had no way down through the burning floors of dense choking smoke. Without any hope of rescue from the roof, as many as two hundred people chose to jump. The people in the Twin Towers on the floors below, where the planes hit, rushed down the escalators. Most made it out. Many had to pass the remains of those who had fallen to their death.

I was sitting at my desk at my office in Chantilly, Virginia. Twenty miles west of the Pentagon. I might have seen the plane on its final descent had I looked out my window.

Lillian, my wife, wasn't so lucky. Looking up from her desk, on the thirteenth floor of one of the tallest buildings close to the Pentagon, in Crystal City, Virginia, she saw Flight 77 hit. First, the entire side of the building exploded, then a wall of fire, then a mushroom cloud billowed over the Pentagon—our Pentagon, the headquarters of the largest, most powerful military in the history of the world. Or so we thought.

The sequence of events from that day escapes me. After twenty-plus years, just a few facts remain that haunt my memory. I remember calling Lillian after the first plane hit the tower in New York. When the second plane hit, my coworkers and I were glued to the television in our conference room.

A friend cried out, "That was planned. There'll be more."

I called from the only unsecured line I could find. I supposed many folks were calling their spouses, children, and parents. Lillian's building was directly in Flight 77's path. Another plane was not an imaginary threat. I told her to leave and get home immediately.

"I can't."

For the next ten hours, I remember being mad at her—really mad. What I didn't realize or appreciate were Lillian's responsibilities. She was the senior employee of her company in the building that day. Her boss and her boss's boss were at the Pentagon meeting with some admirals about the next upgrade to the cruise missile program. She was left behind, working on the software they were all bragging about.

As I drove home, our son Jason called. He was in the first year of college and knew I'd be worried. Then, our daughter Jennifer called. She was living in Los Angeles.

They have always been good children, mostly.

Seeing a commercial airliner crash into a building—the Pentagon, no less—changes you. In an instant, Lillian transformed from an engineer into the building's fire marshal. All federal buildings in DC were ordered to evacuate. Her building, a commercial high-rise, was an afterthought.

Moving slowly, methodically, and deliberately, she looked in every room from the fifteenth to the thirteenth floor, her company's entire workspace. Her words never wavered or changed.

"Time to leave. You must leave now."

When the employees in the simulation lab questioned her authority, her words were simple and direct. When a conference room full of naval officers, including captains and commanders, said they'd be staying, she turned off the power.

I didn't see her for ten hours. The traffic leaving DC was as bad as in one of those disaster movies. I was still mad at her. Perhaps I still am.

At Arlington National Cemetery, there is a memorial to the 184 people who died at the Pentagon that Tuesday: 125 workers from the Pentagon, all 53 of Flight 77's passengers, and 6 crewmembers. The five sides of the memorial honor the five individuals whose remains were never found. The memorial includes each victim's name.

The 9/11 Pentagon Memorial Marker is in Section 64. It's a little over a mile walk. Take a left out of the visitor center on Eisenhower Drive, another left on Sergeant Alvin C. York Drive, and then a right on Marshall Drive. At Patton Circle, take the first exit into Section 64. The memorial overlooks the Pentagon.

Flight 77 lasted eighty-seven minutes. The walk will take you less than twenty-five.

9/11 marker with the Pentagon in the background

MORALE

Captain Miller, next to a letter from home,
your band is the greatest morale booster
in the European Theater.

—General James H. Doolittle

CHAPTER 17

THE DAY THE MUSIC DIED

Gravesite of Major Alton Glenn Miller

THE DAY WAS FEBRUARY 3, 1959. THE BUSES WERE COLD AND cramped. One of the musicians had frostbite, and most had a cold or the flu. It was the tour bus from hell. The Winter Dance Party across the Midwest was in full swing. Buddy Holly was the main attraction with his band of Waylon Jennings, Tommy Allsup, and Carl Bunch. Buddy had left the original members of the Crickets a year earlier due to their insane tour schedule. When their last performance at the Surf Ballroom in Clear Lake, Iowa, ended, everyone just wanted to move on. Ritchie Valens and J. P. "the Big Bopper" Richardson were rising stars and also on the tour. Dion and the Belmonts rounded out what must have been one of the most memorable concert tours in rock-and-roll history.

The bus ride to Moorhead, Minnesota, took the rest of the night and most of the following day. Fed up with being cold, Buddy Holly, just twenty-two and at the height of his career, chartered a small private plane. J. P. Richardson was the oldest of the trio at twenty-nine. He had the flu, and the thought of a bus ride through the freezing Iowa snow was not his first choice. Waylon Jennings gave him his seat on the plane. Tommy Allsup lost his seat to Ritchie Valens in a coin toss. Ritchie's remark, "This is the first thing I ever won in my life," haunts Tommy's memory.[1] The Beechcraft Bonanza had a single engine. The pilot, Roger Peterson, was not instrument-rated and had little experience. It was Iowa in the depth of winter. Their destination was the Fargo, North Dakota, airport—the closest landing strip to Moorhead. Shortly after takeoff, the plane went down. All four on board died instantly.

Learning of her husband's death, the traumatized María Elena Holly had a miscarriage. They'd been married only six months. She blamed herself for his death. She was two weeks pregnant and didn't feel well. Had she gone on the tour with Buddy, she said, "He'd have never got on that plane." She was too upset to attend his funeral.[2]

Rock-and-roll was all the rage. Buddy Holly was the first to introduce the traditional rock-and-roll band of two guitars, a bass, and a set of drums. Four young lads from Liverpool known as the Beatles studied his music and style. They were so taken by Buddy Holly and the Crickets that they used the insect theme to name their own band.

Ritchie Valens was instrumental in popularizing Chicano or Latino rock. He turned the Mexican folk song "La Bamba" into a Billboard Top 40 hit. His ballad to his high school sweetheart, "Donna," rose to become the fourteenth-highest-grossing song of 1959. He was just seventeen years old when the plane crashed.

Don McLean's 1971 song "American Pie" memorialized the iconic phrase "the day the music died." The eight-minute, forty-two-second-long ballad mourned the loss of Buddy Holly, J. P. Richardson, and Ritchie Valens and ended the innocence of early rock-and-roll.

The tour continued when the buses rolled into Moorhead. Because he knew all the words to Buddy Holly's songs, fifteen-year-old Bobby Vee took over lead vocals, and Jennings and Allsup continued with the band for a few more weeks.

Neither Buddy Holly, Ritchie Valens, nor J. P. Richardson served in the military. Buddy and J. P. are buried in their hometowns in Texas. Ritchie Valens is buried alongside his mother at the San Fernando Mission Cemetery in Southern California.

The year 1959 was a hopeful time in America. The Korean War was over, and Vietnam was too new to be noticed. Early rock-and-roll optimism lessened any problems we faced as a nation.

Though it's a bit of a stretch, there is a connection to Arlington National Cemetery. If you spend time walking the grounds, you will see it.

Not everyone buried at Arlington is a war hero—or at least not in the Audie Murphy sense. Walking away from the visitor center and into a distant area, you realize you're walking through ordinary America. Those brave Soldiers were your neighbors, cousins, aunts, and uncles. Some buried at Arlington had extraordinary talent, and others were like you or me. They were from your hometown. Those Soldiers are us—well, the better of us, that is, our better angels. A generation before Holly, Valens, and Richardson, America was engaged in the war to end all wars (or so we thought). In World War II, it felt like every nation was at war.

Alton Glenn Miller didn't need to volunteer. He was thirty-eight and at the height of his career. But the Army needed him much more than he needed them.

Miller was born on March 1, 1904, in Clarinda, Iowa. He attended grade school in North Platte, Nebraska. By eleven, Miller had saved enough money to buy his first trombone. During his childhood, his family moved several times before they settled in Fort Morgan, Colorado. In 1918, he went to Fort Morgan High School and played football. His team, the Maroons, won the Northern Colorado American Football Conference in 1920. He was named Best Left End in Colorado. Before he graduated, he formed his first dance band and vowed to be a professional musician.

In 1923, he entered the University of Colorado in Boulder. While there, he spent most of his time playing gigs and going to auditions. After failing three of his first five classes, he dropped out.

At college, he met Helen Burger, whom he married in 1928. She devoted herself to helping grow his career. Her faithful support and encouragement propelled him to become one of the most successful musicians of the twentieth century.

Miller was the best-selling recording artist of the thirties and forties. Along with playing the trombone, he was a composer, songwriter, and big-band leader. His recordings include "In the Mood," "Moonlight Serenade," "Pennsylvania 6-5000," "Chattanooga Choo Choo," "A String of Pearls," "At Last, I've Got a Gal In Kalamazoo," "American Patrol," "Tuxedo Junction," "Elmer's Tune," "Little Brown Jug," and "Anvil Chorus." Between 1939 and 1942, he had sixteen number-one records. Sixty-nine hits made it into the top ten. Glenn Miller had more top-ten hits than the Beatles or Elvis Presley.

In 1942, at the height of his career, he decided to enlist. He was too old to be drafted, and the Navy didn't want him. Instead, the Army commissioned him a captain in the Army Air Corps with orders to form a band to entertain troops. Stationed in London, his fifty-piece band gave over eight hundred concerts. He continued to record new arrangements at Abbey Road Studios. For a time, he worked with the actor David Niven, a lieutenant colonel in the British Army. Then, in 1944, Miller was promoted to major.

Alton Glenn Miller entertained a generation of young Americans engaged in the global struggle for freedom. General Jimmy Doolittle said, "Captain Miller, next to a letter from home, your band is the greatest morale booster in the European Theater."[3]

In one of his BBC broadcasts, Miller offered, "America means freedom, and there's no expression of freedom quite so sincere as music."[4]

After the Allied invasion of Europe and the liberation of France, Miller received orders to relocate to Paris. On December 15, 1944, he decided to go early to make arrangements for his men; his band was scheduled to fly on December 18. The single-engine plane carrying Major Miller went down over the English Channel. The aircraft and the crew were never seen again. After his death, he was awarded the Bronze Star.[5]

Bronze Star Citation

Major Alton Glenn Miller (Army Serial No. 0505273), Air Corps, United States Army, for meritorious service in connection with military operations as Commander of the Army Air Force Band (Special), from July 9, 1944, to December 15, 1944. Major Miller, through excellent judgment and professional skill, conspicuously blended the abilities of the outstanding musicians comprising the group into a harmonious orchestra whose noteworthy contribution to the morale of the armed forces has been little less than sensational. Major Miller constantly sought to increase the services rendered by his organization, and it was through him that the band was ordered to Paris to give this excellent entertainment to as many troops as possible. His superior accomplishments are highly commendable and reflect the highest credit upon himself and the armed forces of the United States.

Major Alton Glenn Miller's cenotaph memorial is at Arlington National Cemetery, in Section 13. To visit his memorial, walk straight from the visitor center on the paved walkway until Roosevelt Drive, turn right, then turn right again on Wilson Drive.

Look for his grave on a gentle hill. The walk is about three-fourths of a mile and will take you about as long as it does to listen to his three Grammy Award–winning songs, "Moonlight Serenade," "Chattanooga Choo Choo," and "In the Mood."

December 15, 1944—the day the music died, the first time.

Hollywood's Walk of Fame star for Glenn Miller

CHAPTER 18

THE SWEET SCIENCE

Front of gravesite for Technical Sergeant Joe Louis

I WAS IN THE NINTH GRADE IN 1971 WHEN JANIS JOPLIN RELEASED "Me and Bobby McGee." When you're all of fourteen and meet the girl of your dreams, schoolwork is just a necessary evil. Vietnam, hippies, and a fractured country were so common it all seemed normal. The San Fernando Valley community we lived in was about as racially diverse as Iowa. During the Watts riots in downtown Los Angeles five years in the past, thirty-four people died, and over one thousand were injured. Watts, a mostly African American community, is thirty-five miles from our house. Nobody I knew or heard of ever made the forty-minute drive. As a teenager living in the 'burbs in the 1970s, I thought the Apollo program and moon landings seemed closer.

You can get straight C's and still have enough time left over to listen to Janis singing "Me and Bobby McGee."[1] You even had time to learn that Kris Kristofferson wrote the song. Back then, we didn't have YouTube or Spotify. You either had to buy Janis's *Pearl* album, which for me was impossible, or wait your turn on 93 KHJ radio. Fortunately for my young lost soul, the Real Don Steele, a KHJ all-star DJ, knew what I needed to hear. In my head, "Me" was on an endless vinyl loop.

Roger Miller was the first to record the song in 1969. But it was Janis Joplin's death from a heroin overdose and the occasional playing of the song performed by the writer Kris Kristofferson that captivated me. While Janis seemed a bit too radical and dangerous for me, Kris's music and his rough voice spoke to all the troubles everyone I knew could imagine. I can't remember the teacher who explained to me that Kris was a Rhodes scholar and an Army Ranger. His father was a two-star general in the Air Force. Kris flew helicopters, but the Army wanted him to teach literature at West Point. He was the consummate American hero. He turned down all of this to move to Nashville and write songs. For me, though,

Kris's crowning achievement was not being smarter than everyone or even serving his country. His glory came in the ring; he was a Golden Gloves boxer. Anyone can go to Oxford, I thought. But to qualify for the Golden Gloves, surviving ten bouts—*that* was an accomplishment.

As Muhammad Ali once said, "Champions aren't made in the gyms. Champions are made from something deep down inside them—a desire, a dream, a vision."[2]

Not everyone likes boxing. This sport has something visceral about it and is not appropriate for our civilized society. But we like football, the fights in hockey, and yes, at times, boxing. We love the sheer artistry and choreography of a good boxing match. I remember America falling in love with Muhammad Ali. We knew him in his youth as Cassius Clay. We watched him win the light heavyweight gold medal in the 1960 Rome Olympics. Every young boy in the late fifties and early sixties knew Cassius Clay was the 1956 Golden Gloves champion. Not all of America loved Muhammad Ali; I thought they did, but ours was a fractured country back then.

We watched as Cassius converted to Islam and changed his name. We saw him when he refused his draft notice and went to jail. He said he was a conscientious objector. We believed him. Americans believed in this six-foot-three African American. America loved Muhammad Ali. He was a uniter, or so I thought.

By nature, boxing is not a sport used to unite our country. It can be, but often boxing has just the right amount of controlled violence to summon our hate. Professional boxing divided our country at the turn of the twentieth century. In 1910, a boxer named Jack Johnson defeated James Jeffries to defend his heavyweight championship. It was called "the fight of the century." Jim Jeffries was white, while Jack Johnson was unapologetically black. After the fight, race riots broke out across America. For the next decade and a half, white America would send dozens of white boxers to take back their rightful title. Author Jack London called this charge "The great white hope."[3]

Jack Johnson was born into poverty in Galveston, Texas. The son of former slaves, he grew into one of the most influential yet racially divisive African Americans of the early twentieth century. Determined to live free, he moved to New York at age sixteen, where he roomed with Joe Walcott, a welterweight fighter from Barbados. From Joe, all of five feet one, he learned the *sweet science*. He honed his orthodox stance and used an out-boxer style, preferring to stay outside his opponent's reach. Jack Johnson's flamboyant lifestyle made him the focus of hatred throughout white America. In the ring, he lorded over his white challengers, taunting them. Outside the ring, he dared to marry a white woman. America saw Jack Johnson as a divider, fueling the flames of America's original sin.

In 2023, we sit with the luxury of history but still in a fractured country. We long for someone to unite us. Jack Johnson in 1910 and Muhammad Ali in 1970 did as much to divide us as they did to unite us.

At Arlington, though, rests a boxer who united all Americans. Joe Louis Barrow, who went by Joe Louis, was a national hero for everyone. Mothers would tell their children, "If Joe Louis could do it, so can you." He was a better boxer than Johnson or Ali, better than Frasier or Tyson, better than them all. For twelve years, Louis defended his title. In his professional career, he won sixty-six fights and lost only three. Fifty-two were won by knockout. He never lorded over his opponents like Jack Johnson did. He defended his heavyweight title twenty-five times. No one in the history of boxing has ever matched his record, not in any weight class.

Born to sharecropper parents in Alabama, Louis was always quiet and reserved. At age two, his father was committed to an insane asylum, and his mother moved the family to Detroit. When Louis was fifteen, his mother wanted him to play the violin. He had to hide his boxing gloves in the instrument case. A younger boy, who would go on to become another great boxer, Sugar Ray Robinson, reminisced in his autobiography that he idolized his neighbor Joe Louis and used to carry his gym bag.

Louis's career was one for the record books. As an amateur, he competed in fifty-four bouts, losing only four. Forty-three of his amateur fights were won by knockout. He was a Golden Gloves champion three times over.

Determined to shake the racial stigma of Jack Johnson, Louis's managers instilled in him four cardinal rules to live by: "Never have your picture taken with a white woman, never gloat over a fallen opponent, never engage in a fixed fight, and live and fight clean." These rules defined Joe Louis in and outside the ring.[4]

In 1935, the world was descending into war. Everyone knew it was just a matter of time for America to join in. Fascist Italy had invaded Ethiopia, and Louis's fight against Mussolini's six-feet-six Primo Carnera was seen as the vindication of the free world. The six-feet-one Joe Louis knocked the giant out in six rounds.

In 1936, Louis again represented the free world when he fought Max Schmeling. Schmeling symbolized the Nazi Party and Adolf Hitler. Louis thought Schmeling was over the hill. At thirty, Schmeling was long past his prime. Instead of training, Louis took up golf. On the other hand, Schmeling used films of Louis to study his techniques and learn his weaknesses. Both fighters used a boxer-puncher style, making them unpredictable in the ring. Schmeling observed Joe dropping his right after a punch, exposing his head. Schmeling knocked Joe out in the twelfth round.

Upon learning the news of Louis's defeat, noted poet and literary great Langston Hughes observed, "I walked down Seventh Avenue and saw grown men weeping like children, and women sitting in the curbs with their heads in their hands. All across the country that night, when the news came that Joe was knocked out, people cried."[5]

In 1937, Louis fought heavyweight champion James Braddock, knocking him out in eight rounds. Louis refused to call himself the champion, claiming that to gain the title, he first must vindicate his loss to Schmeling. Louis trained seriously for the rematch. He gave up golf. President Roosevelt implored him to win this for the good guys. The pressure on Louis was monumental. In 1938, exactly one year to the day, Louis and Schmeling met again in the ring. Seventy thousand people crowded into a sold-out Yankee Stadium, and seventy million more Americans listened in by radio. Another one hundred million listened in from around the world. Clark Gable, Douglas Fairbanks, Gary Cooper, Gregory Peck, and J. Edgar Hoover sat ringside.

Two minutes and ten seconds into round one, and it was all over. Louis had delivered forty-one punches to Schmeling's two. Three times Schmeling went down. The third time his manager threw in the towel. After the fight, Schmeling was admitted to the hospital with several cracked vertebrae in his back.

In the 1940s, many African Americans questioned serving in the racially segregated military. Louis thought otherwise, stating, "Lots of things are wrong with America, but Hitler ain't going to fix them."[6] Louis, the world's undisputed heavyweight champion, enlisted in the US Army in 1942. Louis never saw combat but was a significant force in recruiting more African Americans to join. He toured throughout the United States and Europe, giving ninety-six exhibition bouts. He helped raise millions of dollars for the war effort. He endorsed the checks from the money he earned over to the Army or Navy relief funds. Over two million Soldiers saw Louis box. He was widely considered America's national hero. In 1945, he was promoted to technical sergeant and awarded a Legion of Merit for "incalculable contribution to the general morale."[7] Rarely given to enlisted men, this prestigious medal qualified him for immediate release from active duty.

Not everyone saw Louis's big heart. The Internal Revenue Service came after Louis for failing to pay the taxes he owed on all his charity bouts. Even though he endorsed the checks over to the Army or Navy relief funds, the IRS considered this income. In his life, Joe Louis would amass over a million dollars in debt to the IRS. The interest alone would total over $50,000 a year.

Joe Louis Barrow died on April 12, 1981. Long before his death, Ronald Reagan, as a part-time actor and radio announcer, toured along with Joe for the USO. As president, Ronald Reagan waived the burial requirements at Arlington, and Joe Louis Barrow was buried with full military honors.

To visit Joe's grave, follow the crowds of visitors to the Tomb of the Unknowns. Stand in silence and respect. After the changing of the guard, exit to the right and follow even more visitors down a small hill. Louis is buried in Section 7A, gravesite number 177. To pay your respects, you must wait your turn.

When you watch the nightly news, you see America as it is. When you learn the story of Joe Louis, you envision a country that could be.

Back of gravesite for Technical Sergeant Joe Louis

AD ASTRA PER ASPERA

We choose to go to the Moon. We choose to go to the Moon in this decade and do the other things, not because they are easy, but because they are hard.

—President John F. Kennedy

CHAPTER 19

A ROUGH ROAD LEADS TO THE STARS

The crew of *Apollo 1* (by NASA/photographer unknown - NASA Images, public domain)

IT WAS JUST LIKE THE THOUSANDS OF OTHER TESTS THE ASTRO-nauts of *Apollo 1* had conducted. For this one, the command module and its three-man crew sat atop an unfueled Saturn rocket on Complex 34 at the Kennedy Space Center. It was late January, and their scheduled ride into space was less than a month away.

Command Pilot Virgil Ivan "Gus" Grissom was one of the seven in the first group of astronauts, the original *Mercury 7.* His suborbital flight in 1961 made him the second American in space. Upon landing in the ocean, the hatch to his *Mercury* capsule prematurely opened. He almost drowned, and his capsule sank. In

Gus Grissom and *Liberty Bell 7* (by NASA/photographer unknown, NASA Images, public domain)

1965, Grissom went on to pilot the first manned Gemini mission. His *Gemini 3* spacecraft was aptly named the *Molly Brown*, after the Broadway musical *The Unsinkable Molly Brown. Apollo 1* would be his third trip into space.

Like Command Pilot Grissom, *Apollo 1*'s Senior Pilot, Ed White, was a space flight veteran. A member of the second group of astronauts called the Next Nine, Edward Higgins White II flew on *Gemini 4* and was the first American to walk in space. White was so excited to be outside his *Gemini* capsule and walking in space that Houston's control center had a hard time getting him back in.

Deeply religious, Ed White carried a small gold cross, a Saint Christopher's medal, and a Star of David into space. Having grown up as an "army brat," White moved a lot as a teenager. Because of his family's frequent reassignments, he didn't have enough time in any one congressional district to secure an endorsement to a military academy. Therefore, as a young high school student, White traveled to Washington, DC, and knocked on several representatives' doors. He eventually convinced Congressman Ross Rizley from Oklahoma to nominate him, thus securing his appointment and admission to the US Military Academy at West Point.

The third member of the *Apollo 1* crew had yet to venture into space. Roger Bruce Chaffee was in NASA's third group of astronauts and a seasoned pilot and test pilot. He graduated from Purdue University and attended the Air Force Institute of Technology in Dayton, Ohio. After graduating from Purdue, Chaffee was commissioned as an ensign in the United States Navy. In October 1962, he was awarded an Air Medal for his actions flying reconnaissance missions during the Cuban missile crisis. Had Roger Chaffee survived the fire and launched a month later as planned, he would

Edward White during *Gemini 4* (NASA Photograph by James McDivitt NASA Images, public domain)

have been the youngest person in space—a record he would have held for over fifty years.

Growing up, all three *Apollo 1* astronauts were Boy Scouts. Chaffee earned Eagle Scout, and all three astronauts married their high school or college sweethearts. They excelled in math and science and were standouts in sports. White missed qualifying for the 400-meter hurdles on the 1952 US Olympic Team by less than half a second. In 1945, Grissom enlisted in the Army Air Corps while still in high school. After West Point, White was commissioned as a second lieutenant in the US Air Force and went on to pilot training. All three pursued advanced degrees and test pilot school to increase their chances of joining the astronaut program.[1]

White, Grissom, and Chaffee were born to be astronauts. As a five-year-old, I watched Air Force Captain Gus Grissom become the second American in space. As a nine-year-old, I taped two lunch thermoses together and fixed a yardstick to the top. I had a hand maneuvering unit as good as Captain Ed White's when he became the first American to walk in space. Add an old football helmet and a barely used baseball glove, and I was ready for my own spacewalk. Standing on the curb, balanced on one foot, I gave

Rocket boys (*left to right*): Gary Hilton, me, and my brother Lowell

a small hop and was free from the capsule. Hissing noises were optional.

When you lost your baseball glove, and I always did, you could blame it on Captain Ed White, who lost his extra glove when they opened the hatch for the spacewalk.[2]

Who needs an extra glove anyway? Playing right field in Little League means having a glove is more for show. Even the best players couldn't hit the ball into the outfield, let alone right field. It might as well be a space glove. I remember that as well.

But then I don't. I don't remember much after 1967, when I turned eleven, at least for a few years after, anyway. Dreams were on hold because of the fire. Apollo was on hold. Ed White, Gus Grissom, and Roger Chaffee were dead.

It was a routine test. It was supposed to be safe. The Saturn stood on the launchpad unfueled. Strapped into their seats aboard their Apollo command module sat the three astronauts. They started down the checklist with the hatch locked from the outside. There was always a checklist. Pilots, test pilots, and astronauts have plenty of checklists. But the fire wasn't on the list.

It started with a voltage surge and then an electrical short. The pure oxygen and positive pressure in the capsule fed the flames. By the time the ground crew could open the hatch, all three men were dead. *Apollo 1* was gone, and the Apollo program was on hold. Also on hold were the dreams of every young boy in America.[3]

United States Air Force Lieutenant Colonel Edward White is buried at his alma mater, West Point. An Air Force officer buried along the Army's *Long Gray Line.* USAF Lieutenant Colonel Virgil Grissom and US Navy Lieutenant Commander Roger Chaffee are buried next to each other at Arlington National Cemetery.[4] Visiting their graves involves walking a little over a mile from the visitor center. Go past the Memorial Amphitheater and the Tomb of the Unknowns. The astronauts are buried in Section 23.

Astronauts and service members Gus Grissom, Ed White, and Roger Chaffee lost their lives in service to our country.

Apollo 1 Memorial at the Kennedy Space Center

CHAPTER 20

HIGH FLIGHT

The Space Shuttle *Challenger* Memorial

John Gillespie Magee Jr. was born in Shanghai, China, to missionary parents. His father was American, and his mother was British. In the early 1930s, they returned to England from China so John could attend Saint Clare College, a preparatory school in Kent, and then Rugby School in Warwickshire. Even though Rugby School was known as the birthplace of the sport of rugby, John excelled at writing poetry. He won the school's first poetry prize in 1938. In 1940, he traveled to the United States, intending to attend Yale. He never made it, choosing the Royal Canadian Air Force instead. Magee was born to fly. In flight school, it took him only six flight hours to solo. The usual time for a cadet's first solo flight was nine to eleven hours. On his first combat mission over the English Channel, his plane was the only one of his fighter group to return. He went on to fly three more combat missions. Then, on December 11, 1941, Magee died in a midair collision with a student pilot who was in his tenth week of active duty. His poem "High Flight" was published after his death.[1]

Oh! I have slipped the surly bonds of Earth
And danced the skies on laughter-silvered wings;
Sunward I've climbed, and joined the tumbling mirth
of sun-split clouds,—and done a hundred things
You have not dreamed of—wheeled and soared and swung
High in the sunlit silence. Hov'ring there,
I've chased the shouting wind along, and flung
My eager craft through footless halls of air . . . .

Up, up the long, delirious, burning blue
I've topped the wind-swept heights with easy grace
Where never lark nor ever eagle flew—

And, while with silent lifting mind I've trod
The high trespassed sanctity of space,
Put out my hand, and touched the face of God.
—John Gillespie Magee Jr.

After the Apollo moon landings, America lost interest in space. Unmanned probes studied the universe but garnered little public attention. The American public was interested only in the manned exploration of space. Faced with low public interest and extreme budget pressures, NASA developed the Space Transportation System, better known as the Space Shuttle. The shuttle program operated from 1981 until 2011 with five orbiter vehicles: *Columbia, Challenger, Discovery, Atlantis*, and *Endeavour*. Each vehicle's six-to-seven-person crew flew to low Earth orbit, deployed satellites, and conducted science experiments. A total of 135 missions were flown, carrying 852 astronauts into space. Numerous shuttle astronauts flew multiple missions; individuals from sixteen countries traveled over 537,114,016 miles.

Designed to carry commercial, Department of Defense, and scientific payloads into space, the Space Shuttle program was supposed to do it all. The cargo bay of *Discovery* deployed the Hubble Space Telescope. The budget needs of the shuttle program changed the entire space industry. Built for launching on unmanned rockets, satellites needed to be modified so they could fit aboard the shuttle. In addition, the Air Force developed its own shuttle crews to deploy classified payloads.

In 1986, I got my first job working for the National Reconnaissance Office. While I didn't appreciate it then, many of the NRO's senior leadership opposed launching our satellites from manned vehicles such as the shuttle. On January 28 of that year, I learned why. Seventy-three seconds into the launch of *Challenger*, it exploded, killing all seven crewmembers. The primary mission of *Challenger*'s flight was to deploy a tracking and data relay satellite (TDRS). This communication satellite was to support NASA and other space organization partners. One of the mission specialists killed was Air Force Lieutenant Colonel Ellison Onizuka. It was Onizuka's job to release TDRS into low Earth orbit.

After the explosion, the NRO's command and control facility in Sunnyvale, California, was renamed in Lieutenant Colonel Onizuka's honor. Our NRO facility, affectionately known as the "Blue Cube," was now Onizuka Air Force Station.

The explosion and loss of all seven astronauts halted the entire space industry. Programs designed to fit aboard the shuttle had to be redesigned to fit unmanned rockets. The Shuttle Launch Complex at Vandenberg Air Force Base was mothballed. Billions of dollars wasted.[2]

While American taxpayers witnessed the loss of some of their money, the American public lost even more. Our nation lost Christa McAuliffe. In 1985, McAuliffe was selected from over eleven thousand applicants to become America's first teacher in space. Schoolchildren from throughout the country and around the world stopped their classes to watch the launch live. McAuliffe planned to give two courses from space. The hopes of millions of children lasted seventy-three seconds. Shortly after the disaster, NASA canceled its teacher program.

A faulty O-ring on one of the solid rocket boosters was the cause of the *Challenger* explosion. As NASA investigated the accident, the shuttle program was reimagined. Commercial satellites lost their rides, and the rework to fit on an unmanned rocket cost private industry billions. It would take thirty-two months before another shuttle would launch.[3]

President Ronald Reagan had planned to deliver his State of the Union address on January 28, 1986. The disaster changed all that. Peggy Noonan, the longtime columnist for the *Wall Street Journal*, was a speechwriter and special assistant to the president when the *Challenger* exploded. Her words, spoken by Reagan, will go down in history as one of the greatest speeches by an American president. Noonan concluded the president's speech with the last line of John Magee's poem written forty-five years earlier. Here are the last few paragraphs of the speech that sought to heal a nation:

> And I want to say something to the schoolchildren of America who were watching the live coverage of the shuttle's takeoff. I know it is hard to understand, but sometimes painful things like this happen. It's all part of the process of exploration and discovery. It's all part of taking a chance and expanding man's horizons. The future doesn't belong to the fainthearted; it belongs to the brave. The *Challenger* crew was pulling us into the future, and we'll continue to follow them.
>
> I've always had great faith in and respect for our space program, and what happened today does nothing to diminish it. We don't hide our space program. We don't keep secrets and cover things up. We do it all up front and in public. That's the way freedom is, and we wouldn't change it for a minute.
>
> We'll continue our quest in space. There will be more shuttle flights and more shuttle crews and, yes, more volunteers,

more civilians, more teachers in space. Nothing ends here; our hopes and our journeys continue.

I want to add that I wish I could talk to every man and woman who works for NASA or who worked on this mission and tell them: Your dedication and professionalism have moved and impressed us for decades. And we know of your anguish. We share it.

There's a coincidence today. On this day 390 years ago, the great explorer Sir Francis Drake died aboard ship off the coast of Panama. In his lifetime the great frontiers were the oceans, and an historian later said, "He lived by the sea, died on it, and was buried in it." Well, today we can say of the *Challenger* crew: Their dedication was, like Drake's, complete.

The crew of the space shuttle *Challenger* honored us by the manner in which they lived their lives. We will never forget them, nor the last time we saw them, this morning, as they prepared for their journey and waved goodbye and "slipped the surly bonds of earth to touch the face of God."[4]

On January 16, 2003, the shuttle *Columbia* launched from Kennedy Space Center in Florida. It was its twenty-eighth and final flight. At launch plus 81.7 seconds, a two-by-one-foot piece of foam broke off the external tank and struck the orbiter's left wing. The collision damaged some of the carbon reentry heat shield tiles. Neither mission control nor the crew noticed. NASA's Intercenter Photo Working Group didn't find the damage until two days into the mission. None of the launch cameras caught a clear view of the damage. The working group chair went to the Shuttle Launch Integration Manager to request images of the damaged tiles. NASA and its two major shuttle contractors, Boeing and United Space Alliance, formed a Debris Assessment Team.

It was common for foam to separate from the external tank during launch. Of seventy-nine missions with clear imagery, sixty-five showed foam strikes occurring. This problem was well known.

NASA's models and simulations told them that the tiles were unlikely to be damaged enough to cause a problem. NASA management decided not to tell the crew. What difference would it make?

On February 1, 2003, the space shuttle *Columbia* burned up as it reentered Earth's atmosphere. People in Texas and Louisiana could see the trail in the morning sky. All seven astronauts lost their lives.[5]

At Arlington, monuments to *Challenger* and *Columbia* are in Section 46, directly behind the Memorial Amphitheater. To visit them, it is a little over three-quarters of a mile. Head straight from the visitor center along Roosevelt Drive, turn right on Farragut Drive and then left on Memorial Drive.

Also buried at Arlington's Section 46 are three members of the *Columbia* crew: Captain David Brown, US Navy; Captain Laurel Blair Salton Clark, MD, US Navy; and Lieutenant Colonel Michael P. Anderson, US Air Force.

The Space Shuttle *Columbia* Memorial

CHAPTER 21

SOMEONE YOU KNOW

Thirty funerals a day (photo courtesy of Daniele Rose)

To experience Arlington, don't just visit. Find someone buried there. It can be a relative, neighbor, or friend of a friend. Perhaps you'll need a history book, but we can all find someone we knew or a loved one who knew someone buried at Arlington. Walking the rolling hills, passing hundreds of our nation's fallen, is to understand our country's most hallowed ground. When you do this and find your person, you'll cease being just a visitor. Instead, our nation's sacred ground for remembrance will become your personal place of reflection.

The winds pick up in autumn, and Virginia's gray skies turn cold. Covered by trench coats, the 3rd Infantry Regiment's dress blues are squared away. With over thirty funerals a day, Arlington reminds you that silence and respect are required. Most burials are simple affairs—a bugler plays "Taps," an honor guard stands at attention, a flag is folded and "On behalf of a grateful nation . . ." is relayed to loved ones of the fallen. You'll sometimes hear a rifle salute, three volleys from a firing party of three to seven riflemen.

Occasionally, you'll see a military funeral with an escort. First, a military band arrives. The size depends on the individual's rank or if the deceased was killed in action. Then slowly, the escort is led by a marching element from the service member's branch. Sometimes, a caisson brings the fallen service member to the final resting place. Here, a six-horse team pulls the two-wheeled artillery caisson carrying the casket. Only the three horses on the left have riders. The horses on the right side are riderless, out of a tradition from earlier days when horses brought supplies and ammunition to the battlefield.

A caparisoned or riderless horse might be added to the procession for US Army and Marine Corps officers of the rank of colonel and above. Here, a member of the Old Guard follows the caisson leading a single horse. The horse wears a saddle, and a Soldier's boots are placed backward in the stirrups, signifying the service member's last ride.[1]

It was cold the day I witnessed one military funeral, complete with a caisson and a caparisoned horse. I heard the band in the distance over a small hill. The musicians were in place, but the processional had yet to arrive. Still a half hour or so before the burial, honor guards were at the ready. With no official party in sight, I approached a young Soldier acting as a road guard.

"How's it going?"

"Well, sir. How are you today?"

"Just trying to understand this place, Soldier."

"Me too."

"How long in the Old Guard?"

"About a year."

"Do you like it?"

"Most of the time, but I'd rather be with the guys I signed up with."

"Where are they?"

"Mostly deployed, sir, Iraq."

"Didn't you volunteer for this?"

"Of course, sir. They told me to."

I watched from a distance. It wasn't my place to interfere. I took no photographs. First came a black limousine. An Army chaplain and a lone widow got out. She was greeted by another formally dressed woman, who got up from a gathering of a dozen or more neatly aligned empty white chairs. I thought it could be a daughter or a friend, but I didn't know. I've since learned that the other woman was one of the Arlington Ladies, an organization that was founded in 1948 by Gladys Vandenberg, the wife of

Air Force General Hoyt Vandenberg. The Vandenbergs routinely attended funerals at Arlington when he was the Air Force Chief of Staff.

The general and Gladys often noticed only a chaplain attended some funerals. Gladys vowed to ensure that no service member would be buried alone without someone from the Air Force family attending. Shortly after the Air Force, the Officers' Wives from the Army and the Navy followed suit. The Marines had no need for a tradition such as the Arlington Ladies. A representative of the commandant's staff attends every Marine funeral at Arlington. If you are a Marine, you already know that.

Although I couldn't be sure, I think the widow had outlived her husband and her entire family. On that cold fall day, just me, an old, retired Air Force officer, and thirty to forty of the widow's Army family bore witness.

While walking among the green grass and hundreds of headstones, you may recall a heated debate over the wisdom of invading Iraq and searching for weapons of mass destruction. Maybe you're old enough to remember the draft and how fractured our country was during the Vietnam War. As you seek your person, walking the manicured lanes, you'll begin to get it. Those kitchen table debates over the wisdom of what this country does or doesn't do become insignificant. As time goes on, those heated discussions with neighbors, colleagues, and maybe that crazy uncle will fade—it's just wasted breath among fools.

As you walk the lanes of Arlington, looking for your person among all those heroes, you will discover what's most important in life—memories such as why your neighbor mowed his lawn on Wednesday nights, often by flashlight. Before, a thousand people around you and none knew the answer. Suddenly, you do. You know because you knew your neighbor. He coached soccer. The matches lasted all day on Saturday. It wasn't polite to mow on Sundays. That left Wednesday nights. Walking at Arlington, you remember what's important.

I'm both fortunate and honored to have known a few people buried at Arlington. I highlighted a boss of mine from years past in a previous chapter. Unfortunately, my youth and his rank meant I didn't know him well. What I do know of Major General Nathan Lindsay comes from friends of mine and, most recently, his wife, Shirley. Her telling me a few precious memories of Nate bonded Shirley and me in a way I never imagined.

Closer to my age and rank are two colleagues who fill my memories with pride, sadness, and hope. These two Airmen, Bill and Mark, provide a sense of the caliber and stature of more common heroes filling Arlington's 639 acres. The following stories are unlike the eulogies offered as we bid them farewell. These memories of mine can't even provide an understanding of the type of person either was. Was Bill a good husband? I think so, but only because I saw Christine cry. Was Mark a good father to Tyler and Matthew? I'm sure he was because he was the type of dad you'd see mowing the lawn on Wednesday nights.

What follows are just some memories. A few of my thoughts that rambled around in my head as I walked to their graves. William Neyman and Mark Roosma were husbands and fathers taken too young. They were also two of the finest officers our nation has ever had.

Bill just showed up at work one day. I didn't know William Neyman long or well. My boss at the time, Colonel Alex, as we called him, never mentioned the guy who would help me. Alex seldom let anyone know what his plans were.

"How's it going?" I said to Bill.

"Excuse me?" Bill responded.

"I mean, what are you doing?" I asked.

"My job. Alex told me to help you."

"Did Alex say I needed help?"

"No. But he said I should ask."

"Okay. I'm Gary. Nice to meet you."

"Bill, happy to help."

"I'm sure."

Bill showed up unexpectedly just as I was trying to restore critical communications to our production center. We were in the middle of the worst power outage we'd ever suffered. My real bosses and Alex were furious, and I needed to find a fix. We traced the data loss to a fiber optic cable break someplace about a mile south of our facility. Our buildings were on the north side of the eighteen-hole golf course at Fort Belvoir, Virginia. Working on an Army base is a bit unusual, and we couldn't just walk around the golf course asking a bunch of questions either.

"Do you need a set of clubs as well?" the guy in the clubhouse asked.

"Nope, just the cart."

Such was the beginning of Gary and Bill's excellent adventure. Unfortunately, Bill had a habit of fading in and out. One second, he was kind, thoughtful, and articulate. The next, it seemed like he was downloading instructions from the mother ship. During those times, Bill would stare off into space with an intensity I couldn't understand. Despite this, we finally traced the cable break to a junction box someplace between the twelfth and thirteenth tees. We found it not because we carried some specialized fiber cable break detector gadget but because we noticed the work crew of twelve or so maintenance workers standing around a backhoe.

Bill wanted a closer look, while I had enough info and needed to get back. Bill jumped out of the cart.

"We can't talk to them, Bill!" I shouted.

"Just be a minute," Bill said, walking toward the work crew.

"Crap." I drove the cart closer so I could hear better.

"Yup, dug up the cable with this backhoe," the crew told Bill.

"Going to have it fixed soon?"

"Within the hour."

"Who are you Air Force guys driving around in a golf cart?"

"We're from ZETA," Bill said.

"Okay."

"ZETA, Bill? Brilliant. Are you from Mars?"

In the evening, after all the day-shifters had left, I confronted Alex. Then it all made sense. Major William Neyman was reassigned to our little unit in Northern Virginia to be near Walter Reed Medical Center and undergo chemotherapy. Bill had brain cancer. Thanks a lot, Alex; I was enjoying not liking this guy. Then you ruined it. For me, there was plenty to admire about Bill. When Bill was focused and attentive, he could size up a complex situation, understand the technical and interpersonal dynamics, and determine the best course of action, including five viable alternatives. I could easily see why the leadership of our organization promoted him to major three years early. This is the fastest Air Force promotion possible. I was a senior captain, and I was to be promoted that year. Bill had less time in the Air Force than I had as a captain. I'd been a captain so long my mother thought it was my first name.

I drove Bill to his chemo treatments at Walter Reed once or twice. This happened a long time ago, and it wasn't as fun as driving around Fort Belvoir in a golf cart. Bill was medically retired two months after I met him. Our small Air Force unit, on an Army base, scrambled to piece together our seldom-worn dress blues. Major William Neyman's funeral was modest but well attended—color guards, "Taps," the folded flag.

I remember the one and only time I saw Bill's wife, Christine. She was walking out of one of the small chapels dotting the rolling hills of Arlington. Her sobs were uncontrollable. Her pain complete. Her whole life changed, marked, and forever damaged. I'd never seen anyone cry like that. I didn't stop and introduce myself. What in the world would I say? I can still see and feel her pain. It's going on thirty-plus years now.

Gravesite of Major William D. Neyman

To visit Major William Neyman's gravesite, turn left out of the visitor center and walk a little less than a mile until you reach Section 8. About three-fourths into the section, turn right. Bill's grave is on a gentle hill surrounded by other good Soldiers.

Little did I know in 1990 I would go through this ordeal again. Fast-forward to 2006 or so, and another young officer would impress me, lighten my spirits, and ultimately sadden my memories with his passing. Suppose the Air Force officer cadre were a baseball team. Mark Roosma would be an A-level major league shortstop. He'd also be a designated batter and the coach. As an Air Force officer, Mark could do it all, make it look easy, and teach others to bat five hundred.

Mark and I shared an office in the newly established Systems Operations Office of the NRO. Long since retired from active duty, I was a senior consultant, an engineering and technical contractor. Sound fancy? Most of the time, I was just a therapist. I say *therapist* because the organization was brand new, and the different groups making up our new organization had hated each other for a long while. My primary job was to keep the senior leadership chewing gum and walking in the same direction. Looking back, I was marginally successful.

Mark was too new to the NRO to understand the history; he was neutral, like Switzerland. Mark's job was to be a representative to NRO headquarters. Mark's station commander was Colonel Mike R. Mike was a prince; Mark had it easy. If that wasn't enough to make for a great assignment, Mark's good friend was the operations office's director, Colonel George B., who was even nicer than Mike. Their families were good friends, but George and Mark kept this on the down-low. Had I known it then, I wouldn't have cared or been interested; my focus was on the other site commanders—you know, the kind of bosses with Type A personalities. I had a miserable job. Mark had a happy-go-lucky confident personality. I had long since grown into a cynic, and during this job, I had stopped liking anyone. I had nothing to lose and tried to bring Mark down with me. I failed.

Once, Mark and Bob, another site representative, decided to run the Marine Corps Marathon. Mark was a runner, kept himself fit, and didn't have an extra ounce of unnecessary body fat. Bob was just beginning to show that all too common middle-aged spread. Bob's site commander was—let's just say—challenging. I often went to lunch with Bob. Because of his boss, he often needed therapy. Perhaps I should have been a priest. As the years pass, I can't remember ever inviting Mr. Happy Mark to lunch. But I remember lunches with Bob and that the fries at Five Guys come from Rigby, Idaho.

I asked about their success on the Monday after the big weekend race. Bob said he'd never run so far or for so long in his life. Mark simply replied, "Me either." Turns out, after finishing the entire twenty-six miles of the marathon, Mark turned around and ran back to find Bob. I could never get a straight answer from either of them, just how many miles Mark ran that day. Are you beginning to see why I didn't always like Mark being happy?

This other story is cloudy due to the gallon or so of beer I'd consumed. I'm not a big beer drinker, but it becomes downright delicious somewhere around the fourth one. The commanders and the headquarters' leadership had gathered for some technical design meetings at an off-site.

Half of my job was facilitating the operations organization's strategic planning, where everyone confessed their hopes and dreams. This proved extremely important as our entire organization was completely reorganized. My other half was carving up the NRO's operations budget. I can't divulge the total dollar amount, but let's just say there were plenty of zeros to make all the station commanders happy.

It was six in the evening; I'd had too much beer and needed food. As the commanders and their site representatives poured themselves into a fleet of rental cars, I had a brilliant idea. Why don't we stop and scale the sheer cliff adjacent to our hotel? "Excellent idea," was the slurred response from the crew inside my vehicle. The cliff was a rock face of about a hundred feet straight up. Given the jagged nature of the rock, the climb would be a technical challenge for even the healthiest among us.

First out of the car, I charged up the hill. I made it about seven feet straight up before falling to the ground. The next thing I knew, commanders, site reps, and support staff were all charging up the hill and trying to find suitable hand- and footholds. One by one, colonels and captains fell back to earth. Then somebody started yelling, "Stop, Mark, stop!" Luckily, Mark hadn't consumed nearly the amount of beer most of us had. Somehow, not just one person but all of us realized Mark could actually do it. Mark could scale the entire rock, all one hundred feet straight up. Mark was in good enough shape and, at that point, had more coordination than sense. "Stop, Mark, stop!" we all yelled as Mark conquered the summit. And there you have it—Mark Roosma, king of the hill and master of all he surveyed.

Bill Neyman was thirty-two when he passed. Mark Roosma made it to fifty-six before cancer caught up with him. Christine Neyman remarried many years later. Life goes on. You've got to admire a man who marries a woman with a big piece of her heart missing.

I missed Mark's funeral. We moved to California to care for my parents, and Lillian's mother was in and out of consciousness. She was in her last few months. I wanted to go to Mark's service; I tried. Knowing Mark, it was a big party. Family and friends formed a prayer group a year or so before he passed. We got regular updates on Mark's progress. For a while, we all believed Mark would beat it. The pictures we routinely got of Mark, his sons Tyler and Matthew, and his wife, Meaghan, appeared happy. In every single photograph, Mark was smiling. The wonderful pictures and the favorable prognosis told us of bravery and courage.

I'm sure this was true for them both. There was a time for Bill and Mark when they could see their cancers in the rearview mirror. But bravery and courage get you only so far. A cheerful, upbeat demeanor often hides a family's fear of the loneliness that most certainly follows.

A few years after Mark's funeral, I realized I was glad I missed it. I saw Christine cry years ago; that was enough. I'm not a brave person. I'm often not happy. I saw Christine cry. Seeing Meaghan cry would have been too much.

Major William Neyman and Lieutenant Colonel Mark Roosma were guardians of the high ground. We lost them before they could become part of the new United States Space Force. They continue to watch over us.

William Neyman's grave is a little less than a mile walk. Take a sharp left onto Eisenhower Drive from the visitor center. Continue to Section 8. Just before you reach Patton Drive, his grave, number 9984, is on the right.

To visit Mark's grave, take an immediate left out of the visitor center on Eisenhower Drive and walk past Section 54. Turn left on Leahy Road. Mark's grave is about halfway down Section 55 on the right. He is in the same section as his father, Colonel John S. Roosma Jr., and his uncle, Major General William A. Roosma, who is in Section 54. The family requested that they all be buried near each other, which is a comfort to Mark's mom and aunt.

This chapter is for Christine, Meaghan, and Shirley Lindsay, and all the wives whose husbands rest at Arlington.

Gravesite of Lieutenant Colonel Mark D. Roosma

CLOSING THOUGHTS

Privates next to colonels

The mystic chords of memory, stretching from every battlefield and patriot grave to every heart and hearthstone, all over this broad land, will yet swell the chorus of the Union, when again touched, as surely they will be, by the better angels of our nature.

—Abraham Lincoln, first inaugural address, March 4, 1861

By the time Lincoln delivered these words, seven states had already seceded from the Union. The first shots would be fired from Fort Sumter in a little over a month, beginning the Civil War.

America is much more than the heroes buried at Arlington National Cemetery. But through these heroes, we can see beyond the troubles of today that divide us.

Why did Audie Murphy stand on a burning tank fending off dozens of enemy soldiers? Why did John Basilone return to the fight only to die so young? Why did a young airman jump on an ignited magnesium flare and crawl toward an open cargo bay door? Of course, they did these heroic acts of bravery for their fellow comrades. But did they also act with selflessness and sacrifice simply for us? Didn't they also act fearlessly for those who would inherit the country they fought to save?

There was no reason for Glenn Miller to want to serve his country. He was a superstar, an entertainer with limitless potential. After Joe Louis defended his heavyweight title thirteen times, why did he enlist as a private?

If you stand beside any grave at Arlington, you will begin to see a story that doesn't start or end on those beautiful rolling Virginia hills. If you stay at Murphy's gravestone long enough, you might see his sharecropper's family from Hunt County in northeast Texas. Out of these twelve children, only the seventh, Audie Murphy, would be buried at Arlington. If you followed him from the battlefield to Hollywood's Walk of Fame, you might find yourself touched by the kindness of his wife, who, after he died, dedicated her life to caring for disabled veterans.

Presidents William Howard Taft and John Fitzgerald Kennedy are buried at Arlington, as are five of our country's five-star generals and admirals. Also interred there is John J. Pershing, the only person to be promoted to General of the Armies—six stars during his lifetime. Even George Washington had to wait until America's bicentennial in 1976 for Congress to present his sixth star.

But Arlington is not just for the famed or noteworthy. Our nation's most hallowed ground is also for the sharecropper's son, the emancipated slave, and the Confederate soldier who fought so gallantly. Walking the gentle hills overlooking our nation's capital, you will see colonels resting beside privates. You will find surgeons and nurses beside the troops they tried to save. As you tour the cemetery's grounds looking for the famous, you must pass by thousands of graves known only to history.

As I walk the lanes of Arlington, I find comfort and solace. I've come to realize that our national place of reflection and remembrance is just a waypoint guiding us to what is good about this great land. The stories about the men and women buried at Arlington do not begin or end on those Virginia hills. The four hundred thousand stories of Arlington make the search for America's better angels a little clearer. If you allow it, Arlington can be your lens to help focus your resolve.

As you conclude your visit to America's venerable ground, you are reminded that these most honored dead are not as unique as our history books might suggest. Most notably, these heroes are all of us. They are from our hometowns and cities. They reflect us at our best and instill in us a recognition of our failings and hopes. At Arlington, it is easy to find a hero. You don't need to walk far. The hero you seek lies just the next grave over.

As you return home, your vision is sharper. Your friends and neighbors take on a new light; they reflect faint glimmers of momentary acts of kindness and sometimes beacons shining on lifetimes of service and devotion. As you seek your better angels, you might find them closer than you think. They have been next to you all along.

Until our next walk. As President Lincoln said in his Gettysburg Address,

> That from these honored dead we take increased devotion to that cause for which they gave the last full measure of devotion—that we here highly resolve that these dead shall not have died in vain—that this nation, under God, shall have a new birth of freedom—and that government of the people, by the people, for the people, shall not perish from the earth.

Not forgotten

ACKNOWLEDGMENTS

PORTIONS OF THIS BOOK, INCLUDING CHAPTER 5, ARE BASED on material from the Congressional Medal of Honor Society. This nonprofit organization was chartered by Congress in 1958 to preserve the Medal of Honor's legacy. As the nation's highest military award for valor, the medal recognizes a servicemember's actions of bravery and heroism in the face of extreme risk of life. CMOHS promotes the values of the Medal of Honor, which include courage, sacrifice, integrity, commitment, patriotism, and good citizenship.

I greatly appreciate the time and attention afforded to me by the CMOHS executive director, John Falkenbury, and executive assistant, Karen Turpin. They are truly heroes serving heroes.

This book would not be possible without the help of a few folks who edited my writing and helped get the manuscript ready for publication. Each of these talented reviewers brought something valuable and different to the project. First, Sheila edited my drafts, laughed at my mistakes, and made suggestions on clarity and impact. Sheila is unapologetically English and drives around Southern California with a bobblehead of the queen (now the king) in her rear window. Second, Sandra is the editor of *Dispatches* magazine, the quarterly publication of the Military Writers Society of America. Sandra is an ex-military spouse and knows full well the inequities of life in the military. Sandra published a few of these chapters in *Dispatches* and pushed me to finish the manuscript. Sharon pulled together my disjointed thoughts and stories into the book you see here. Sheila, Sandra, and Sharon worked harder than me and are smarter than me, and this effort proves it.

Through it all, my eternal partner, Lillian, has held the light as we explored the world. Her encouragement, edits, and comments are my North Star.

NOTES

Introduction

1. "Letter to Mrs. Bixby," Abraham Lincoln Online, from *Collected Works of Abraham Lincoln*, edited by Roy P. Basler et al., accessed November 13, 2023, www.abrahamlincolnonline.org/lincoln/speeches/bixby.htm.

Chapter 1: Memorial Day

1. "Memorial Day History," US Department of Veterans Affairs, National Cemetery Administration, accessed September 7, 2023, www.cem.va.gov/history/Memorial-Day-History.asp.

2. "Casualty Status," US Department of Defense, August 21, 2023, www.defense.gov/casualty.pdf.

Chapter 2: Arlington House

1. "Architecture and Construction," Arlington House, the Robert E. Lee Memorial, National Park Service, last updated July 23, 202, www.nps.gov/arho/learn/historyculture/architecture.htm.

2. "Robert E. Lee and Slavery," National Park Service, last updated May 10, 2022, www.nps.gov/arho/learn/historyculture/robert-e-lee-and-slavery.htm; "Robert Edward Lee," National Park Service, last updated January 18, 2022, https://www.nps.gov/arho/learn/historyculture/robert-lee.htm.

3. Elizabeth Pryor, "Robert E. Lee (1807–1870)," Encyclopedia Virginia, Virginia Humanities, December 7, 2020, encyclopedia virginia.org/entries/lee-robert-e-1807-1870/.

4. John Kelly, "Md. Bridge History Includes Breach That Couldn't Be Spanned," *Washington Post*, April 21, 2010, www.washingtonpost .com/wp-dyn/content/article/2010/04/20/AR2010042004795.html.

5. William C. Dickinson, "Gen. Meigs and Arlington National Cemetery," *Washington Post*, June 23, 1997, www.washingtonpost.com /archive/opinions/1997/06/23/gen-meigs-and-arlington-national -cemetery/2fa30f89-e8b4-4d5e-8920-215540c8edcf/.

Chapter 3: The Good Son

1. "The Lincoln Family," National Park Service, last updated May 10, 2021, www.nps.gov/liho/learn/historyculture/the-lincoln-family .htm.

2. "Robert Todd Lincoln at Harvard College," Abraham Lincoln Online, accessed April 10, 2023, www.abrahamlincolnonline.org /lincoln/sites/harvard.htm.

3. "Fort Sumter," American Battlefield Trust, accessed April 10, 2023, www.battlefields.org/learn/civil-war/battles/fort-sumter.

4. "Family: Robert Todd Lincoln (1843-1926)," Mr. Lincoln's White House, Lehrman Institute, accessed April 10, 2023, www .mrlincolnswhitehouse.org/residents-visitors/family/family-robert -todd-lincoln-1843-1926/.

5. "The Death of Willie Lincoln," Abraham Lincoln Online, accessed April 10, 2023, www.abrahamlincolnonline.org/lincoln /education/williedeath.htm.

6. Joe Servis, "The Surrender Meeting," National Park Service, last updated June 14, 2022, www.nps.gov/apco/learn/historyculture/the -surrender-meeting.htm.

7. Todd Arrington, "Robert Todd Lincoln and Presidential Assassinations," July 2014, National Park Service, accessed April 10, 2023, www.nps.gov/articles/000/robert-todd-lincoln-and-presidential -assassinations-not-formal-title.htm.

8. "Wills House Virtual Identity: Thomas 'Tad' Lincoln," National Park Service, October 19, 2021, www.nps.gov/gett/learn/historyculture/wills-house-virtual-identity-thomas-tad-lincoln.htm.

9. Encyclopedia.com, s.v. "Mary Todd Lincoln," last updated May 18, 2018, www.encyclopedia.com/people/history/us-history-biographies/mary-todd-lincoln.

10. "The Short Life of Abraham Lincoln II," Lincoln Collection, July 28, 2014, lincolncollection.tumblr.com/post/93136038279/the-short-life-of-abraham-lincoln-ii; "Abraham Lincoln II," Alchetron, last updated October 7, 2022, alchetron.com/Abraham-Lincoln-II.

11. "The Pullman Company," Pullman History, last updated April 2020, www.pullman-museum.org/theCompany/.

12. Jason Emerson, "This Man Was the Only Eyewitness to the Deaths of Both Lincoln and Garfield," *Smithsonian Magazine*, January 7, 2022, www.smithsonianmag.com/history/read-an-exclusive-excerpt-from-the-diary-of-the-only-person-present-at-the-deaths-of-both-lincoln-and-garfield-180979296/; Arrington, "Robert Todd Lincoln."

13. Philip Jett, "Robert Todd Lincoln—Blessed and Cursed," History Reader, August 7, 2018, www.thehistoryreader.com/historical-figures/robert-todd-lincoln-blessed-and-cursed/.

14. Arrington, "Robert Todd Lincoln."

15. "Robert Todd Lincoln Tomb," Abraham Lincoln Online, accessed April 11, 2023, www.abrahamlincolnonline.org/lincoln/sites/robert.htm.

Chapter 5: The Medal of Honor

1. "Hershel Woody Williams," Woody Williams Foundation, accessed September 7, 2023, woodywilliams.org/woody-williams-bio.html.

2. "Chief Warrant Officer 4 Hershel Woodrow Williams, USMCR," Marine Corps University, accessed September 7, 2023, www.usmcu.edu/Research/Marine-Corps-History-Division/Information-for-Units/Medal-of-Honor-Recipients-By-Unit/Cpl-Hershel-Woodrow-Williams/.

3. Medal of Honor, U.S. Code 10 '8291 (2018).

4. "Military Awards for Valor - Top 3," US Department of Defense, accessed August 25, 2023, https://valor.defense.gov/.

5. "Statistics & FAQs," Congressional Medal of Honor Society, accessed August 25, 2023, https://www.cmohs.org/medal/faqs.

6. "A History of Heroism," Congressional Medal of Honor Society, accessed August 25, 2023, https://www.cmohs.org/medal/timeline.

7. Congressional Medal of Honor Society, home page, accessed September 7, 2023, www.cmohs.org.

8. "Medal of Honor (MOH) Recipients at Arlington National Cemetery," Arlington National Cemetery, accessed September 7, 2023, www.arlingtoncemetery.mil/Explore/Notable-Graves/Medal-of-Honor-Recipients.

9. Congressional Medal of Honor Society, *The Medal of Honor: 160 Years of Courage and Sacrifice* (Clearwater, FL: PMG, 2023). You may download the book at https://www.dropbox.com/s/l9yvch6rrr6qcat/MOH%20-%20Full%20Book.pdf?dl=0. Through nationwide outreach and educational programs, Medal of Honor recipients speak directly to community and veteran groups, fostering a legacy of honor and integrity. They share their stories, inspiring young people and adults to practice positive values and leadership in their own communities. CMOHS facilitates a museum and a library, sharing its resources with the public to preserve the significance of our nation's highest medal and its importance in history. Every year, CMOHS honors civilians from throughout the country for acts of heroism and service.

Chapter 6: Jimmy Doolittle

1. "General James Harold Doolittle," United States Air Force, accessed September 7, 2023, www.af.mil/About-Us/Biographies/Display/Article/107225/general-james-harold-doolittle/.

2. "History: The Doolittle Raid," Children of the Doolittle Raiders, accessed November, 13, 2023, www.childrenofthedoolittleraiders.com/doolittle-raiders-history/.

3. Carroll V. Glines, "Mama Joe's Tablecloth," *Air & Space Forces Magazine*, August 1, 1989, www.airandspaceforces.com/article /0889mamajoes/.

Chapter 7: Tip of the Spear

1. Alex Perry, "Inside the Battle at Qala-I-Jangi," *Time*, December 5, 2001, content.time.com/time/subscriber/article/0,33009,1001390,00 .html.

2. "CIA Memorial Wall," Central Intelligence Agency, accessed November 13, 2023, https://www.cia.gov/legacy/headquarters/cia -memorial-wall/.

3. Douglas Waller, *Wild Bill Donovan: The Spymaster Who Created the OSS and Modern American Espionage* (Free Press, 2011), 12.

4. "William J. Donovan," Department of Justice, Criminal Division, accessed August 10, 2016, www.justice.gov/criminal/history/assistant -attorneys-general/william-j-donovan.

5. Thomas A. Rumer, *The American Legion: An Official History, 1919–1989* (New York: M. Evans, 1990), 107.

6. Ronald H. Spector, *In the Ruins of Empire: The Japanese Surrender and the Battle for Postwar Asia* (New York: Random 2007), 8.

Chapter 8: Twenty Seconds

1. John T. Correll, "20 Seconds Over Long Binh," *Air & Space Forces Magazine*, April 1, 2005, www.airandspaceforces.com/article /0405levitow/.

2. Howard E. Halvorsen, "Air Force History: John Levitow and *Spooky* 71," *Tinker Air Force Base*, March 3, 2017, https://www .tinker.af.mil/News/Article-Display/Article/1103646/air-force -history-john-levitow-and-spooky-71/.

3. "Airman First Class John L. Levitow," Air Mobility Command Museum, accessed September 7, 2023, amcmuseum.org/history /airman-first-class-john-l-levitow/.

4. C. Douglas Sterner, "John Levitow: Nightmare on *Spooky 71*," Home of Heroes: Medal of Honor & Military History, accessed September 7, 2023, homeofheroes.com/heroes-stories/vietnam-war/john-levitow/.

Chapter 9: Call Me Nate

1. "Major General Nathan J. Lindsay," United States Air Force, last modified February 1992, www.af.mil/About-Us/Biographies/Display/Article/106391/major-general-nathan-j-lindsay/.

Chapter 10: Mama Joe's Tablecloth

1. "Tablecloth, Gen. James H. Doolittle," Smithsonian National Air and Space Museum, accessed September 7, 2023, airandspace.si.edu/collection-objects/tablecloth-gen-james-h-doolittle/nasm_A19740535000.

Chapter 11: Pamela Opal Lee

1. Audie Murphy, *To Hell and Back* (New York: Henry Holt, 1949).
2. "Pamela Opal Lee 'Pam' Archer Murphy," Find a Grave, accessed November 13, 2023, www.findagrave.com/memorial/51117433/pamela-opal_lee-murphy.
3. "America's Most Decorated World War II Combat Soldier," Audie L. Murphy Memorial Website, accessed September 7, 2023, www.audiemurphy.com.
4. Dennis McCarthy, "Remembering Shy Heroine Pam Murphy on Veterans Day," *Los Angeles Daily News*, August 28, 2017, www.dailynews.com/2014/11/06/dennis-mccarthy-remembering-shy-heroine-pam-murphy-on-veterans-day/.

Chapter 12: A Blank Check

1. "Audie Murphy," Arlington National Cemetery, accessed September 7, 2023, www.arlingtoncemetery.mil/Explore/Notable

-Graves/Medal-of-Honor-Recipients/World-War-II-MoH-recipients /Audie-Murphy.

2. Jack Cheevers, "Sepulveda VA Hospital to Be Torn Down," *Los Angeles Times*, March 15, 1994, www.latimes.com/archives/la-xpm -1994-03-15-mn-34353-story.html.

3. Lindsay, "The VA Sepulveda Ambulatory Care Center aka the American Embassy from 'Argo,'" *I Am Not a Stalker* (blog), January 17, 2013, www.iamnotastalker.com/2013/01/17/the-va-sepulveda -ambulatory-care-center-aka-the-american-embassy-from-argo/.

Chapter 13: Lena

1. Michael Robert Patterson, "John Basilone — Gunnery Sargeant, United States Marine Corps," Arlington National Cemetery, accessed November 13, 2023, www.arlingtoncemetery.net/johnbasi.htm.

2. "John Basilone Biography," USS Basilone Association, accessed September 7, 2023, ussbasilone.org/john-basilone-biography/.

3. Joseph A. Grasso, *Manila John: The Life and Combat Actions of Marine Gunnery Sergeant John Basilone, Hero of Guadalcanal and Iwo Jima* (Pittsburgh, PA: Rose Dog Books, 2010), 193.

4. Grasso, *Manila John*, 193.

5. Teagan Fredericks, "Great Love Happens Once: The Enduring Story of John and Lena Basilone," United States Marine Corps, February 14, 2019, www.1stmardiv.marines.mil/News/News-Article -Display/Article/1758035/great-love-happens-once-the-enduring -story-of-john-and-lena-basilone/.

Chapter 14: Medgar and Myrlie

1. "Medgar Evers: July 2, 1925-June 12, 1963," Wesleyan University, June 24, 2013, www.wesleyan.edu/mlk/posters/pdfs/evers.pdf.

2. Myrlie Evers-Williams and Manning Marable, *The Autobiography of Medgar Evers: A Hero's Life and Legacy Revealed through His Writings, Letters, and Speeches* (New York: Basic Civitas Books, 2005).

3. Jordan Ginder, "Biographies: Medgar W. Evers," National Museum of the United States Army, accessed September 23, 2023, www.thenmusa.org/biographies/medgar-w-evers/.

4. *Ghosts of Mississippi*, directed by Rob Reiner (Columbia Pictures and Castle Rock Entertainment, 1996).

Chapter 15: It's Not Just a Job

1. Norman Friedman, *US Battleships: An Illustrated Design History* (Annapolis, MD: Naval Institute Press, 1985).

2. US Surgeon General, "Special Appendix" in *Report of the Surgeon-General, U.S. Navy, 1898* (Washington, DC: Government Printing Office, 1896), 173.

3. Patrick McSherry, "Battleship U.S.S. *Maine*," Spanish American War Centennial Website, accessed September 7, 2023, www.spanamwar.com/maine.htm.

Chapter 16: Another Day That Will Live in Infamy

1. Nick Routley, "A Timeline of Everything That Happened on 9/11," World Economic Forum, September 10, 2021, www.weforum.org/agenda/2021/09/9-11-timeline-visualized-america-september-terror-attacks/.

Chapter 17: The Day the Music Died

1. Paul Galloway, "Hit Parade," *Chicago Tribune*, June 24, 1988, https://www.chicagotribune.com/news/ct-xpm-1988-06-24-8801100141-story.html.

2. Wendy Mead, "The Day the Music Died: Rock's Great Tragedy," Biography, September 8, 2020, www.biography.com/musicians/the-day-the-music-died—plane-crash.

3. Michael E. Ruane, "75 Years Ago, Glenn Miller Vanished on a Flight over the English Channel," *Washington Post*, December 24, 2019, washingtonpost.com/history/2019/12/24/glenn-miller-is-missing-years-ago-big-band-mega-star-vanished-flight-over-english-channel/.

4. Collin Makamson, "'Jukebox Saturday Night': Glenn Miller's Last Hit as a Civilian," National WWII Museum New Orleans, December 5, 2012, nww2m.com/2012/12/jukebox-saturday-night-glenn-millers-last-hit-as-a-civilian/.

5. *The Glenn Miller Story*, directed by Anthony Mann (Universal International, 1954).

Chapter 18: The Sweet Science

1. "Biography," Kris Kristofferson, accessed August 31, 2023, https://kriskristofferson.com/biography/.

2. "History: When, Where & How Did It All Begin?" Golden Gloves of America, accessed August 31, 2023, www.goldenglovesusa.org/about/.

3. John Ridley, "A True Champion Vs. The 'Great White Hope,'" NPR, July 2, 2010, www.npr.org/2010/07/02/128245468/a-true-champion-vs-the-great-white-hope.

4. Joseph Vincent, "The Impossible Greatness of Joe Louis—Boxing's Most Dominant Champion (The Real Captain America)," Joseph Vincent, YouTube, July 14, 2021, www.youtube.com/watch?v=FK_i12K4g74.

5. Ben Wyatt, "The Legacy of Joe Louis' Loss to Max Schmeling on Juneteenth," *Guardian*, June 19, 2021, www.theguardian.com/sport/2021/jun/19/joe-louis-max-schmeling-heavyweight-fight-boxing-junctccnth.

6. Ira Berkow, "Joe Louis Was There Earlier," *New York Times*, April 22, 1997, www.nytimes.com/1997/04/22/sports/joe-louis-was-there-earlier.html.

7. Ellen Fried, "VIPs in Uniform: A Look at the Military Files of the Famous and Famous-To-Be," *Prologue Magazine*, National Archives, Spring 2005, vol. 38, no. 1, www.archives.gov/publications/prologue/2006/spring/vips-military.html.

Chapter 19: A Rough Road Leads to the Stars

1. Ben Cosgrave, "The Apollo 1 Launchpad Fire: Remembering Grissom, White and Chaffee," *Life*, accessed September 7, 2023, www

.life.com/history/the-apollo-1-launchpad-fire-remembering-grissom-white-and-chaffee/.

2. "Mission Monday: Five Fast Facts about the First American Spacewalk," Space Center Houston, June 1, 2020, spacecenter.org/mission-monday-five-fast-facts-about-the-first-american-spacewalk/.

3. NASA Content Administrator, "Apollo I," NASA, August 7, 2017, www.nasa.gov/mission_pages/apollo/missions/apollo1.html.

4. Elizabeth Howell, "Apollo 1: A Fatal Fire," Space.com, January 26, 2021, www.space.com/17338-apollo-1.html.

Chapter 20: High Flight

1. Bryan Swopes, "18 August 1941: 'High Flight,'" This Day in Aviation, August 18, 2023, www.thisdayinaviation.com/tag/john-gillespie-magee-jr/.

2. Mark C. Cleary, "USAF Space Organizations and Programs: Defense Department Involvement in the Space Shuttle Program," *The Cape: Military Space Operations*, chapter 1, section 3, spp.fas.org/military/program/cape/cape1-3.htm.

3. William P. Rogers et al., *Report to the President by the Presidential Commission on the Space Shuttle Challenger Accident* (Washington, DC: NASA, June 6, 1986), chapter 4, sma.nasa.gov/SignificantIncidents/assets/rogers_commission_report.pdf.

4. Ronald W. Reagan, "Explosion of the Space Shuttle *Challenger* Address to the Nation," NASA, January 28, 1986, history.nasa.gov/reagan12886.html.

5. *Columbia Accident Investigation Board, Report Volume I* (Washington, DC: NASA, August 2003), s3.amazonaws.com/akamai.netstorage/anon.nasa-global/CAIB/CAIB_lowres_full.pdf.

Chapter 21: Someone You Know

1. "Military Honors," Arlington National Cemetery, accessed August 31, 2023, www.arlingtoncemetery.mil/Funerals/Funeral-Information/Military-Honors.

INDEX

ABOUT THE AUTHOR

When you meet someone at age twelve, fall in love, and spend the next fifty years together, it's hard to know where one person starts and the other ends. I retired from the USAF after twenty-two years of service. My bride, Lillian, raised two delightful children and rose through the ranks to become a vice president for a Fortune 500 tech company.

In 2020, my father passed away due to complications from dementia. It's not hereditary, but memories fade even for the best of us. Before they are lost, I plan to finish my memoir, *Aim High, a Memoir*. With these stories, hopefully, my children will learn a little about who I was and not just what I did.

When not on grandpa duty, I blog to entertain and thank those who gave me more than I could ever give back.

Together with Lillian, we've recorded our travels since 2007. Those blog posts to family and friends were quick updates on our whereabouts rather than any serious attempt at stringing together a coherent story. Practice helped improve my prose, but the places we visited became the real story. The pictures chronicled our adventures and were what everyone wanted to see anyway.

People ask me why I decided to write and share my stories in retirement. I responded with the following:

"Retirement is hard enough; why not get paid?"

"In the military, we were always leaving someone behind, so I write to share memories and keep in touch with the people we've befriended."

In my memoirs, I enjoyed reliving the exploits of three of my friends during my military years. None of them are as good as my memories, but therein lies the reality I want to remember. Now that I've written a little about them, my stories became the truth. At least, the truth for me. Isn't life better that way?